SO YOU WANT TO BE AN AIR CONDITIONING CONTRACTOR?

SO YOU WANT TO BE AN AIR CONDITIONING CONTRACTOR?

by

Delbert D. Thomas

Copyright © 2001 by Delbert D. Thomas
All rights reserved. No part of this book may be reproduced, stored in a retrieval system, or transmitted by any means, electronic, mechanical, photocopying, recording, or otherwise, without prior written permission from the author.

ISBN: 1-58721-946-8

This book is printed on acid free paper.

About the Book

Written to assist those who are considering the purchase of or the starting of a heating, ventilating and air-conditioning (HVAC) business. Describes the different categories within this exciting and profitable industry. They include the service and repair of equipment as well as the installation of HVAC systems in new and existing structures. This is probably the first book addressing this matter.

How much capital is required? What equipment will be needed? How are customers located? How are overhead costs calculated? Describes how to prepare an estimate and submit a proposal. Describes the type of work where the cash flow is better. Avoid the pitfalls in bidding jobs. How can payments be assured?

Includes engineering tables to assist in Design/Build projects. Describes the many types of HVAC systems available.

Table of Contents

COMMENTS BY THE AUTHOR 1

Chapter 1
WHAT TYPE OF CONTRACTING? 3

Chapter 2
HOW MUCH MONEY WILL I NEED? 39

Chapter 3
WHAT TYPE OF EQUIPMENT WILL I NEED? 41

Chapter 4
DIFFERENT SYSTEMS TO BE ENCOUNTERED 49

Chapter 5
THE COMPRESSION REFRIGERATION CYCLE 91

Chapter 6
PREPARING YOUR FIRST BID 101

CHAPTER 7
CASH FLOW 109

TABLE 1
ROUND DUCT SIZING 117

TABLE 2
ROUND TO RECTANGULAR DUCT CONVERSION 119

TABLE 3
WATER PIPE SIZING 121

TABLE 4
GALLONS PER FT CONTAINED IN PIPE 123

TABLE 5A
  REFRIGERATION LINE SIZING (R22) 125

TABLE 5B
  REFRIGERATION LINE SIZING (R134a) 127

TABLE 6
  REFRIGERANT PRESSURE / TEMP CHART 129

TABLE 7
  STEAM AND CONDENSATE RETURN LINE SIZING .. 131

TABLE 8
  ELECTRIC MOTOR FULL LOAD AMPS 133

TABLE 9
  ELECTRIC WIRE AMPACITY AT TEMPERATURE RATINGS 135

TABLE 10
  EVAPORATIVE COOLING SYSTEM SIZING 137

TABLE 11
  HANDY FORMULAE 139

TABLE 12
  ENTHALPY (H) VALUES OF DRY AIR 141

TABLE 13
  COOLING LOAD CHECK FIGURES 143

ATTACHMENT "A"
  SAMPLE RESIDENTIAL BID 145

ATTACHMENT "B"
  SAMPLE JOB COST BREAKDOWN 147

SO YOU WANT TO BE AN AIR CONDITIONING CONTRACTOR?

COMMENTS BY THE AUTHOR

After over 45 years in the industry as a mechanical contractor, manufacturer's representative and consulting engineer I have a pretty good idea on what is required to become an air conditioning contractor. Most of those who enter this business have previously been a journeyman, an equipment vendor or worked in some capacity with an air conditioning contractor. Often he (she) has labored under the misconception that as an employee he (she) is the one that is making the boss rich and therefore should be his (her) own boss. Wow, what a concept!

If you take the step of becoming a contractor it will probably be some time before you make as much money as you did when you were just an employee with no worries about making payroll, collections, law suits and labor disputes. This all being true, why would anyone want to become his own boss? There are this who are satisfied with a 40 hour week and a pay check on Friday. Then there are others who want to control their own destiny and are willing to put in the extra effort to obtain this goal.

I have been one of this second group. Starting as a new contractor is a lot like becoming a new parent. You start as a rank amateur and make all of the mistakes in handling your children. There are now many good manuals on how to become a good parent. Unfortunately I have never seen a book offering advise on how to become a successful air conditioning contractor. This book seeks to overcome this lack of information and is based on many years of experiencing what works and just as important, what doesn't work.

If you choose to enter into this exciting business, I wish you the best of success.

Delbert D. Thomas ME
September 1999

Chapter 1

WHAT TYPE OF CONTRACTING?

OK, you want to be a heating, ventilating and air conditioning (HVAC) contractor. There are any number of types of business opportunities to be considered, some of which are outlined as follows:

1) Service and Repair Business

    A) Residential
        a) Air Conditioner & Heating Equipment Repairs
        b) Service Calls & Vehicles
        c) Equipment Replacement

    B) Commercial
        a) Large Chillers
        b) Package Type Equipment
        c) Boilers & Industrial Heat Applications
        d) Service Calls & Vehicles
        e) Equipment Replacement

2) Contracting

    A) Residential Contracting
        a) Custom Homes
        b) Multi-family
        c) Housing Tracts
        d) Air Conditioning Non-Air Conditioned Houses
        e) Marketing & Sales

B) Commercial & Institutional Contracting
   a) High Rise Buildings
   b) Low & Mid Rise Buildings
   c) Schools
   d) Prisons
   e) Hospitals
   f) Marketing & Sales

C) Specialty Contracting
   a) Temperature Controls and Automation Sub Contracting
   b) Piping Sub Contracting
   c) Ductwork Sub Contracting
   d) Marketing and Sales

The choice to be made is somewhat dependent on the past experience of the one who plans on opening his own business. Always start with the type of business where you are most comfortable. If you have been a journeyman sheet metal worker with only experience in the residential area, you certainly have no business tackling hospital projects. Also if you have been a salesman for a commercial equipment supplier you probably don't belong in the custom home market. In the Author's opinion, some type of HVAC engineering background is necessary. The American Society of Heating, Refrigerating and Air Conditioning Engineers (ASHRAE) has an excellent correspondence course available. Some Trade associations also have basic courses available.

It's probable that anyone wanting to go into the air conditioning contracting business has had some association with the industry. Possibly as a journeyman, supplier, estimator, project manager or sales person for a contractor. If you fall into one of the above categories you will probably want to go into the type of work with which you are most familiar.

If you are from outside of the industry, you should align yourself with someone within the industry who is knowledgeable. For the sake of argument, let's say you have a background in the financial industry. This could be of great advantage to you because of your business knowledge. You will however be dependent on someone who knows the sales and marketing required as well as someone who knows the pertinent phase of the construction industry. You should probably consider purchasing a going business or at least the assets thereof. Often a contractor wants to retire and is willing to stay on until the new owner gets is feet on the ground. As with any business acquisition you have to look careful at any possible liabilities. There may be litigation lurking in the wings or some delinquent accounts payable owed to a supply house.

The advantage of acquiring an operating business is that there may be a decent backlog which will help with cash flow. Often the seller places a financial value on "Good Will" or has an impressive list of customers. These "assets" are of questionable monetary value. Customers may leave when the Owner retires and therefore the value of a "Customer List" may not be very great. This is not to say that the list of customers may not be valuable. It all depends on the business relationship at the time of transfer of the ownership.

Adequate capital is required to start a business from scratch so you need to consider this fact. There are tools and equipment to purchase as well as equipping and staffing an office. Trucks will be required as well as sufficient payroll reserve to cover you until your first checks begin to arrive.

Residential work requires the least capital as most all of the material involved can be purchased from suppliers. Most of the ductwork involved will be round and is available from the local supply house. Any minor specialty fabrication can be done by any duct fabricator. There will be a minimum of equipment to purchase. Traditionally the payment requests drawn against the contracts are more prompt than with commercial or institutional work. In the case of housing projects it's possible to negotiate semi monthly progress payments. Payments are usually made within a 30 day period and if you are receiving semi monthly progress payments you should expect these every 15 days. Because many residential general contractors are smaller than the general contractors on large commercial projects, they may also be underfinanced. You should be careful and check out the financial condition and payment history of any general contractor with whom you contemplate entering into a contract.

Light commercial work requires a little more capital as you will only be receiving progress payments one time per month and assuming you have to wait 30 days for the check to arrive you have to be able to carry your payroll and some of your material for 60 to 75 days. Depending on what area of the country you are in there is generally a duct work fabricator you can supply most of your rectangular duct requirements. A minimum amount of sheet metal fabricating equipment is suggested so you can fabricate minor fittings to prevent delays caused by waiting for shipment from the duct fabricator. A list of equipment is provided in the Chapter 3.

If you plan on becoming involved in large commercial,

institutional or hospital work considerable expense must be allocated to equip a sheet metal fabricating shop. There is a considerable amount of good used sheet metal fabricating machinery available from machinery dealers or contractors going out of business. Nevertheless even used equipment is relatively expensive. This type of work also involves piping and steam fitting. Here you have an option. You can set up and operate your own piping department or you can subcontract this phase out to others.

The cost to set up a piping department is not as great as setting up a sheet metal department but the expense still should be considered before taking this step. A list of required equipment is provide in Chapter 3. The Author's opinion is that unless you have a strong background as a pipe fitter, you should start by using a reliable subcontractor. In many large metropolitan areas there are subcontractors who do nothing else but pipe fitting and setting related equipment. In suburban areas there may not be such contractors and this might impact your decision on the types of jobs you undertake. Let's take a look at the different markets and what they entail.

1) SERVICE AND REPAIR BUSINESS

A) RESIDENTIAL SERVICE

Residential service and repairs are a portion of the residential contracting business. This covers the emergency service and repairs of heating and air conditioning systems. In addition to emergency service calls pre-season "tune ups" of heating and cooling equipment can be a good source of revenue. This requires newspaper, radio and television advertisements as well as a direct mail campaign.

Service technicians must be experienced with the type of equipment which they are liable to encounter on service calls. In areas where fuel oil is the predominant fuel the service technician should have a solid knowledge of oil burners and oil handling systems. Likewise in areas where natural gas or propane is the predominant fuel the service technician should have a solid knowledge of gas burners. In either case approximately 90% of all service calls will be electrical or control related. The service technician should be well grounded in electrical theory and be able to read a wiring schematic. In the event that no wiring schematic is available he should be capable of designing a control system which will provide the operational control as well as maintaining all safety features of the system.

a) Air Conditioning & Heating Equipment Repairs

In areas where residential air conditioning is common, the service technician should have a sound basic knowledge of refrigeration. All service technicians must have an EPA license to work on the refrigeration circuits he will encounter. As in the case of heating systems the service technician needs a strong background in electrical theory and control systems as the majority of service calls will be related to electrical or control problems.

Whether the service technician is engaged in heating service, cooling service or both, he should have at least an elementary knowledge of air handling systems and in the case of hot water systems a knowledge of pumps and piping systems. Most manufacturers conduct training schools which are invaluable to the service technician.

b) Service Calls & Vehicles

Each service technician should be assigned a service truck in the form of a van or pick up truck. There is a difference of opinion as to which is best. A pick up truck should have a utility body, with the center of the bed retained open, or the vehicle should be enclosed like a van or a truck with a camper shell. In either case the vehicle should be arranged with bins and compartments to organize the parts and components that may be required on service calls. An inventory list of all parts and components should be maintained. When parts are installed, they should be replaced on the vehicle when it returns to the base. Each service truck should have a price list of all components so the service technician can prepare an invoice at the service location.

Someone at the base location should be assigned the duty of receiving service calls. This individual may be a clerical person or if the size of the operation warrants, a service manager. The calling customer should be advised as to the payment basis required. This should include the hourly rate and whether the payment can be made using a credit card or if cash or a check will be required. Unless the customer has an established account, payment should be collected at the time of service by the service technician. The service company would be wise to obtain a "merchant account" and be able to accept credit cards.

The person receiving the service request call prepares a service work order which contains the name, location, type of equipment and nature of the call. The service call can then be assigned to the service technician best qualified to handle the specific call or one locate in the area where the service call originated. Each truck should be radio dispatched or the service technician equipped with a cellular phone or pager. In this manner, the service tech does not have to return to base when other calls come in.

c) Equipment Replacement

Because the service technician is in a position to inspect existing equipment he is in an ideal position to recommend replacement with equipment of higher efficiency. He therefore is able to add to the sales volume of the contracting department. It is wise for management to develop a commission program based on the amount of equipment sold. In this manner the service technician becomes a member of the Contractors sales force and becomes motivated to sell as well as repair.

B) COMMERCIAL SERVICE

a) Large Chillers

Large commercial and institutional buildings more than often use chilled water systems. There are 4 basic types of chiller used to produce chilled water. Many commercial service companies also specialize in equipment replacement. To that extent they are actually contractors but not in the sense that they bid on construction work. All service technicians must have an EPA license to work on the refrigeration circuits he will encounter. As in the case of residential systems the service technician needs a strong background in electrical theory and control systems as the majority of service calls will be related to electrical or control problems.

Centrifugal Chillers

These machines are generally encountered in sizes of 200 tons and larger although they are available in sizes as small as 100 tons. Centrifugal compressors are relatively simple devices but require special knowledge and training although the training required is not intensive. Service technicians with centrifugal chiller experience are often those who have been employed at some time by one of the major manufacture's service department. In some large metropolitan areas the Local Refrigeration Union may conduct training courses for this type of equipment.

Screw Chillers

Screw compressors have been the mainstay of the industrial refrigeration industry for many years. Only recently have screw compressor chillers become universally acceptable within the air conditioning industry. Chillers as small as 40 tons may be encountered and as large as 1,500 tons. Rarely are screw compressors serviced in the field with the exception of new seals when the compressors are of the open drive type. When major compressor service is required, the compressors are traditionally removed from the chiller and a replacement compressor installed. There are many firms engaged in rebuilding such compressors so the service technician rarely need the expertise to perform a major compressor overhaul.

Reciprocating Chillers

Reciprocating compressors have been the mainstay of the air conditioning industry since it's inception. Chillers are available from 5 ton through 500 tons although when a chiller becomes larger than 200 tons, screw or centrifugal compressors are more commonly encountered. Reciprocating compressors are available as "semi-hermetic" type, where the motor is internal and is cooled by the refrigerant or as "open" type where the motor is external of the compressor. Field service of semi-hermetic compressors is limited to the replacement of valve plates and external components. Field service of open type compressors is limited to the replacement of valve plates, seals, motors and externally attached components. Major overhauls are rarely done in the field as many firms have rebuilt reciprocating compressor available and in stock for replacement and therefore the service technician is not required to have the ability to overhaul or rebuild reciprocating compressors.

Rotary Scroll Compressor Chillers

Many of the newer small chillers (less than 100 tons) are now available with hermetic scroll compressors in lieu of reciprocating compressors. The compressors in these types of chillers are not serviceable and require replacement rather than repairs

Absorption Chillers

Prior to the 1950s the absorption chiller was the mainstay of large tonnage projects. These machines were fired by steam and are still commonly found in facilities which require the use of steam such as hospitals. The most commonly encountered absorption chiller today is the gas fired absorption chiller. These are available from 5 tons through 2,000 tons. The service and repair of absorption chillers is highly specialized and unless the Contractor has personnel that has experience in such equipment he should avoid the service of this equipment. If however the Contractor wishes to pursue this market, there is considerable demand in the large metropolitan areas for service companies who are experienced with absorption chillers. Contact should be made with the local gas utility company who can be a great source of leads.

b) Package Type Equipment

Package equipment is just as it's name implies. The entire cooling, and sometimes heating, plant is contained within a single cabinet. This type of equipment runs the gamut from residential type 2 ton roof top units to 2,000 ton penthouse type units found in large industrial plants. Regardless of the size these units operate on the same principals. The smaller units use either hermetic type or semi-hermetic type reciprocating or rotary scroll type compressors. The large sizes generally use screw type compressors with the largest sizes using multiple compressors.

The hermetic reciprocating and rotary scroll type compressors cannot be serviced in the field. Service on the semi-hermetic reciprocating type compressors is limited to the replacement of valve plates and externally mounted components. Service on the semi-hermetic type screw compressors is limited to replacement of external components. The service technician's expertise on the reciprocating and rotary scroll compressors is on the same level as required for the residential service contractor. The expertise level of the service technician working on the larger machines with screw compressors is higher and of the same level as those servicing the screw compressor type chillers.

c) Boilers & Industrial Heating

Boilers may produce steam, at any given pressure, or hot water. The boilers may be fired by natural gas, propane, fuel oil, electricity or coal. They may be of the cast iron sectional type, fire tube type or water tube type. Commercial and industrial heating covers the emergency service and repairs of heating systems. In addition to emergency service calls pre-season "tune ups" of heating equipment can be a good source of revenue. This requires qualified personal as well as a direct mail campaign.

Service technicians must be experienced with the type of equipment which they are liable to encounter on service calls. In areas where fuel oil is the predominant fuel the service technician should have a solid knowledge of oil burners and oil handling systems. Likewise in areas where natural gas or propane is the predominant fuel the service technician should have a solid knowledge of gas burners. In either case approximately 90% of all service calls will be electrical or control related. The service technician should be well grounded in electrical theory and be able to read a wiring schematic. In the event that no wiring schematic is available he should be capable of designing a control system which will provide the operational control as well as maintaining all safety features of the system.

Service of coal fired equipment should be left to those specialized firms experienced with installing and servicing stokers.

Whether the heating service technician is engaged in steam, hot water or air heating systems, he should have at least an elementary knowledge of air handling systems, of hot water systems and pumps and a basic knowledge of steam theory and the servicing of such items as traps, pressure regulators and other steam specialities. Should the service technician lack experience in any of the above systems, the service call should be referred to others.

d) Service Calls and Vehicles

Each service technician should be assigned a service truck in the form of a van or pick up truck. There is a difference of opinion as to whether the pick up should have a utility body, with the bed retained open, or whether the vehicle should be enclosed like a van or a truck with a camper shell. In either case the vehicle should be arranged with bins and compartments to organize the parts and components that may be required on service calls. An inventory list of all parts and components should be maintained. When parts are installed, they should be replaced on the vehicle when it returns to the base. Each service truck should have a selling price list of all components so the service technician can prepare an invoice at the service location.

Someone at the base location should be assigned the duty of receiving service calls. This individual may be a clerical person or if the size of the operation warrants, a service manager. The calling customer should be advised as to the payment basis required. This should include the hourly rate and whether the payment can be made using a credit card or if cash or a check will be required. Unless the customer has an established account, payment should be collected at the time of service by the service technician. This person receiving the service request call prepares a service work order which contains the name, location, type of equipment and nature of the call. The service call can then be assigned to the service technician best qualified to handle the specific call or one locate in the area where the service call originated. Each truck should be radio dispatched or the service technician equipped with a cellular phone or pager. In this manner, the service tech does not have to return to base when other calls come in.

e) Equipment Replacement

Because the service technician is in a position to inspect existing equipment he is in an ideal position to recommend replacement equipment. He therefore is able to add to the sales volume of the contracting department. It is wise for management to develop a commission program based on the amount of equipment sold. In this manner the service technician becomes a member of the Contractors sales force and becomes motivated to sell as well as repair.

2) CONTRACTING

A) Residential Contracting

a) Custom Homes

This market is probably the simplest to enter. Because it may be the simplest to enter does not mean that basic knowledge of the business is not a major factor. Most projects within this market place are of the Design / Build (D/B) type. This means that the contractor needs to be able to both design and install the heating and cooling systems. This requires that the Contractor have enough engineering back ground that he is able to design a workable system. The Tables at the end of this book provide basic engineering data which will assist the new contractor with his education.

Many custom homes are designed by an Architect who has engaged a Consulting Engineer to design the HVAC systems. More than often this has not been done and possibly only a cooling and heating load has been done on the structure. In such cases the Air Conditioning and heating contractor is expected to do his own layout. The building design often dictates the type of cooling and heating system to be used. If the building has a steep roof or one with tile, a roof mounted package may not be the best application. In such cases a split system would probably be the best choice. This type of system has the condensing unit located on a slab outside of the building with the air handling portion located in an attic, basement or in a closet. If the project is located in an area with natural gas, a gas furnace or boiler will probably be the best choice. If natural gas is not available the choice becomes either fuel oil, propane, electric heat.

When using a fossil fuel such as natural gas, propane or fuel oil the Contractor has to address the products of combustion. The Building Code in the specific area where the installation is to be made will dictate the type and method of removing the products of combustion and the method of providing combustion air to the appliance. The extension of the flue and combustion air ducts must be considered when locating the furnace or boiler. Long nearly horizontal flue runs should be avoided. A good rule of thumb is that the horizontal distance must be no greater than 33% of the vertical height. When this cannot be accomplished, an induced draft fan should be provided. In all cases the local Building Code should be followed.

Special consideration should be given to the use of propane or butane. This gas is heavier than air and a gas leak can cause the gas to settle to the floor and build up until a spark or pilot light is encountered at which moment the roof of the house will be transported to the next county. Special handling of propane and butane is covered in many Building Codes. In same cases the combustion appliance may not be installed in a basement and a drain line from the appliance enclosure must be provide to an area outside the building. In some areas this drain line (actually a duct) may be as large as 8" in diameter.

In general the gas fuels require a double wall flues where the flue passes within 6" of combustible material. These flues are referred to as a Type "B" flue and are constructed with an outer jacket of galvanized steel and an inner tube of aluminum. The final hook up between the appliance and the flue is referred to as the breeching and can generally be constructed of single wall galvanized material which is classified as a Type "C" flue and must be kept 6" or greater distance between the breeching material and any combustible material such as wood. Where fuel oil is used a double wall flue material with refractory insulation between the walls of the flue must be used. This type of flue is referred to as a Type "A" flue.

Because the furnace or boiler requires air to accomplish combustion, air from the outside of the house should be supplied. The amount of combustion air duct and/or opening is generally specified at 2 square inches per 1,000 of fuel input. This amount is equally divided between an area located within 6" of the floor and 6" from the ceiling of the appliance location. In any event the local Building Codes must be consulted and followed when it comes to flues, combustion air and fuel hook up to the appliance.

If electric heat is selected, the most economic choice,

from an operating stand point, would be a heat pump where the same unit provides both cooling and heating. When considering a heat pump, be sure the heating output is sufficient at outside design conditions. If the unit does not provide sufficient heat under these conditions either the size must be increased or auxiliary electric heat must be included in the air stream. The auxiliary electric heat is a good idea as this heat will come on during defrost cycle when cool air is being discharged from the indoor coil.

Heating may either be forced air or hydronic. A forced air heating system uses a furnace, heat pump or an air handler served by a hot water boiler. Steam boilers for single family dwellings are rarely encountered today and therefore are not covered in this book. Hydronic heating uses a hot water boiler to provide the heat which is pumped to heating base board radiation or convectors.

In cold climates hot water radiant heating is a good approach. In this system, copper or cross linked polyethylene tubing is attached to the underside of the flooring or embedded within a concrete floor. The floor is then warmed creating a comfortable condition for the occupants. A drawback to radiant heating is that once the floor mass gets warm, it hold and radiates the heat and in areas with rapid temperature changes occur, the space temperature will overshoot the set point of the thermostat.

Hot water boilers may be of the copper tube, steel tube or cast iron type. In very cold climates the cast iron may be the most efficient however in areas where the temperature can vary 25 degrees to 30 degrees within a few hours a copper tube boiler may be the best choice. The cast iron boiler retains residual heat so that if the temperature of the space varies rapidly, the conditioned space temperature will overshoot the thermostat set point. The copper tube boiler contains much less water and therefore has the ability to heat up or cool down much more rapidly.

Once the method of providing cooling and heating has been established, it is necessary to distribute the cooling and heating to the various areas of the house. This requires the ability to design the duct system or the hot water piping system.

If the distribution system is to be forced air, the furnace, air handler or package air conditioner must be located. If the equipment is to be located within the house or the basement, the location of the flue and combustion air will have an impact on equipment location. Once the equipment has been sized and the location has been selected a duct layout must commence. If the structure is single story and has a basement or crawl space the duct routing is relatively easy. If the climate conditions are such that cooling is the major concern, the registers and/or diffusers should be mounted ain or near the ceiling. If possible the return intake should be located in or near the floor. If however the concern is mainly for heating, floor registers located beneath windows are a good location. In such a case the return intake should be in or near the ceiling.

When the supply ductwork is to be installed within an attic it is important to determine if the roof structure is based on trusses or is a "stacked roof" with ceiling joists and rafters. If trusses are employed, the use of a horizontal furnace is not possible. It is also important that the trusses be laid out where the supply air plenum will clear the trusses because trusses cannot be cut and "headed off." A stacked roof presents no such problems. Depending on the pitch and height of the roof, there will be room for a horizontal furnace or air handler. When selecting an up flow furnace or air handler, the ceiling joists can be cut with the opening framed to permit the supply plenum to penetrate into the attic space. Be sure that any plenums or ducts penetrating into the attic space are not located under a valley near an outside wall. In such cases there will be inadequate room.

In the single story case the furnace or air handler can be either an up flow, down flow or horizontal. If a down flow furnace is placed on a wooden floor, a heat resistant base is generally required. If a horizontal furnace is chosen, it is important that service access be provided as well as a flue arrangement which will provide sufficient draft. This is a particular problem if the horizontal furnace is installed in the attic. Quite often the vertical flue stack is too short to develop a draft. In such cases the flue should be extended or an induced draft fan should be employed. A platform for the furnace and a built in ladder or "pull down" stairs should be provided for access to the horizontal unit as well as providing an electric light fixture near the unit.

A roof mounted unit requires a platform, base or frame so that the unit can be installed level. Both the supply and return ducts from the unit must penetrate the roof into the attic or into a furred space prepared for the ductwork. If the roof is truss construction, the ducts will have to pass between the trusses in which case the duct widths cannot exceed 22" on a standard 24" trussed roof. A stacked roof can have the rafters cut and headed off to make room for the ducts.

Multi-level structures and those with "Cathedral" or open beam ceilings can present a challenge. Ducts must be run in chases and furred down areas to deliver the air to the various rooms on different floors.

In all cases provision must be made to drain the condensate from every cooling coil. If a cooling coil is located on a second floor or above a ceiling, codes quite often require a secondary drain in case the primary drain becomes plugged. Codes sometimes require that this condensate be terminated within a dry well. It is never permissible to run a condensate directly into a plumbing vent or waste line. The condensate may be terminated under a sink or bathroom lavatory into the fixture tail piece above the trap. Be sure local codes are followed in this matter.

In the case of a split system the refrigerant liquid and suction lines must be run in a manner where they cannot be damaged during construction. This is particularly important where the lines run vertically within a wall cavity. Concern over penetration by dry wall screws or lathing nails must be given. Some type of protection from nail holes must be provided. A final item of importance is the location of the air conditioning or heat pump condensing unit. Never locate the unit under a bedroom window if you expect the home owner to get a good night's sleep. A patio installed condensing unit can be a noisy nuisance during a summer barbeque.

The insulation of ductwork and refrigerant lines is important and building codes and State energy requirements should be reviewed. In all cases the refrigerant suction line should be insulated with a foam rubber material at least ½" thick. The joints should be sealed with adhesive or a special rubber tape designed for this purpose. Failure to do this will cause condensate to form on the exterior of the cold line and drip water onto ceilings or within wall cavities.

If the system incorporates air conditioning the duct insulation must have a sealed vapor barrier. Failure to do this will allow moisture to condense on the exterior walls of the ductwork, soak the insulation which destroys it's efficiency and eventually leak onto the ceiling, within the wall cavities or into the basement if such exists. 1" x .5# density insulation should be considered as a bare minimum for air conditioning applications with 2" x .5# preferred by the Author. Systems with heating only do not require the vapor barrier but the thickness and density of the insulation should be maintained.

An alternate material for interior ductwork is fiberglass. Codes permit Class 2 flexible ductwork to be used. Rigid fiberglass (round or rectangular) is also a good option. In all cases the Contractor should follow the manufacturer's installation instructions as well as local building codes. Sealing of joints between sections and to metal fittings are of extreme importance.

When installing flexible fiberglass ductwork, it is important to minimize sagging as such causes air pressure drop. The temptation by the installer to allow surplus flex duct to loop around rather than cutting the flexible duct must be avoided. Allowing this situation causes increased pressure drop which should be avoided. Sharp bends should be avoided for the same reason. The radius of the bend should never be less than the diameter of the duct.

Flexible aluminum duct, both insulated and bare, is available and meets most codes for residential work. The Author does not recommend it's use because of excessive air pressure drop because of the corrugations in the metal. If this material must be used, the Author recommends that the duct size be increased to the next commercially available size to offset the air pressure drop problem.

Ducts exposed to the weather should be internally lined with a minimum of 1" x 1.5# density fiberglass liner with a neoprene facing to prevent air erosion. This liner should be attached with approved adhesive and stick pins and washers which are designed for this purpose. Because the liner is installed on the interior of the duct, the sheet metal duct dimensions must be increased to compensate for the liner.

The Tables at the end of this book will assist in duct sizing.

b) Multi-Family Units

These structures include duplexes, triplexes, condominiums and apartment houses. The work on these units is quite similar to single family dwellings and are more than often constructed for the lower end of the market. Perhaps they are closer in some ways to the lower end of the tract housing market. The exception is for some condominiums which are closer to the mid and upper quality level of custom homes. Package AC units and split systems are the type of air conditioning most commonly encountered. Depending on the area of the country, some lower cost units may be equipped with wall heaters.

Large apartment and condominiums may use central plants for both cooling and heating. Each unit may have a fan coil or small air handler being served with chilled and / or hot water from a central chiller and boiler location. Units with fan coils should have a minimum amount of ductwork as most are equipped with direct fans which are capable of a minimum amount of air pressure drop. Units with air handlers may be treated in a similar manner to forced air furnaces as most are capable of delivering the required volume of air at between .3" and .5" of external static pressure.

c) Housing Tracts

This type of work is very similar to the single family dwelling market with the exception that all of the dwellings are built on a speculative basis. This means that the houses will be constructed before they are sold. You will not be dealing with the ultimate home owner but the developer. The developer wants to construct the houses on the most economical basis consistent with the price range for which the units will be sold. When dealing with the home owner you will find he is amenable to upgrades whereas the developer will be more resistant. Almost without exception tract housing work will be awarded on the lowest first cost. This doesn't mean that relationship with the developer isn't important however it is now pencil sharpening time.

Before the new contractor enters this market he, or his key personnel, should have a background in this highly competitive market. This type of work can be highly profitable providing the contractor has good control of his labor costs, has designed a system which can be installed economically and has a well organized operation. The contractor should review his operation and see if he can handle the production which will be expected. Large tract projects can move at the rate of 5 to 15 houses per day. It is important to have the qualified manpower, tools and equipment to handle this rate of production. Sufficient money for payroll and materials must be available to carry between payment draws.

d) Air Conditioning Non-Air Conditioned Houses

This market is similar to the new single family market with the exception that you will be the "Prime Contractor." This means that there will be no general contractor to do the cutting and framing, no roofer to provide roofing, no electrical or painting contractor to perform their phase of the work. That's right, it's all your responsibility and you will need your own personnel or reliable subcontractors who are capable of performing these phases of the project.

The work will take place in a dwelling that will probably be occupied and consideration for the home owner is important. As little disturbance of the household as possible must take place. Scheduling around rainy weather can be important as you won't want to open holes through the roof if rain is on the way.

The system design is very important as the requirements for furring duct chases and soffits should be kept to a minimum. Here is where creativity is important and the contractors ability in sales is paramount. It's necessary to go over all details with the home owner and obtain his approval of what is to be done.

If it becomes economically impossible to run ductwork to some portion of the dwelling, consider multiple units and if flue vents are a problem consider a small outdoor type boiler and chiller with lines running to multiple fan coil unit to provide either cooling or heating. Each fan coil could be controlled from a separate wall thermostat.

Be skeptical that an existing forced air furnace can

produce sufficient air for air conditioning so some research is required. If the fan in the furnace is direct driven from the motor, plan on a new furnace if a cooling coil is to be attached. It's important to remember that 400 cfm per ton of cooling is required and the furnace should be capable of supplying this amount at .5" static pressure.

e) Marketing & Sales

If you have acquired an existing company engaged in the residential market there may be a customer list including general contractors. It will be advisable to meet as many of these individuals as possible and learn what projects they may be planning. Becoming acquainted with Architects and Building Designers who specialize in residential design is an excellent means of learning about up coming projects. One of the most fruitful sources of new projects will be established local sub contractors such as electrical, concrete, framing and grading. Obviously you don't go to an air conditioning and heating contractor who will be a competitor. This is an excellent method of "networking" and will assist all parties involved.

A plumbing contractor is also a good contact providing that he does not have his own HVAC department. Some areas have combination shops that provide both plumbing and HVAC . They often submit package bids with no "break out" between the two trades. This leaves the independent plumbing contractor at a disadvantage if no competitive heating and air conditioning bids are received. The independent plumbing contractors will be pleased to assist you in identifying new projects. In some cases they may even agree to introduce you to the builder or developer.

Local real estate offices may be a good source. They may have handled the land sale for a development and be willing to provide the proper contact's name. They may be also valuable to you if they know you specialize in adding air conditioning to existing homes. A seller may be in a position to offer to have the house air conditioned if a buyer purchases the house. The real estate agent can provide the seller with this information or possibly the buyer if he plans an providing cooling to the house after purchase.

Another good method is to drive around and learn which general contractors are building the types of units with which you are interested. Go by the general contractor's office and introduce yourself. When you see a piece of ground being graded contact the grading contractor and learn who the developer is. Contacting the developer this early in the game may bring you into a project you may otherwise never have heard about until to late.

Residential HVAC contractors are wise to associate themselves with a national brand of equipment. If you plan on specializing in the single family market select a manufacturer who is known for (or assumed to be noted for) high quality equipment. If you are looking at the volume market you need to associate yourself with a competitive line of equipment that is more suitable for this particular market. This does not mean that the equipment should be of low quality standards just that certain manufacturers concentrate on national advertising where others may save money on national advertising and the cost of their equipment reflects the lower marketing costs.

Most of these companies have a cooperative advertising

plan where they share the cost of advertising. The company may also have been contacted by a large developer to learn who the recommended "dealer" for their products is in your area. It also pays to establish a relationship with the local supply house. Not only is it important to establish a credit line but these companies may also know about projects that are coming out to bid.

Advertising in local "throw away" papers may be also be a good source as well as local radio and cable television spots. This is particularly true if your equipment manufacturer will participate in the advertising costs. Unless you are heavily involved in the service and repair business, advertising in the telephone book "Yellow Pages" is of questionable value. These methods are secondary in importance to the methods listed in earlier paragraphs.

B) Commercial & Institutional Contracting

This type of work is entirely different than in the residential market place. There are rare exceptions where the HVAC contractor is successful in both residential and commercial / institutional markets. There are different types of projects under this grouping. Some fit the HVAC contractor who is more oriented towards the "air side" and as such may be better qualified to follow the package equipment market. Those more oriented towards the "wet side" may be better qualified for central chilled and hot water systems.

a) High Rise Building

Buildings higher than four or five stories might loosely be considered as "High Rise" although to some who are used to 40 story buildings the five story structure may seem quite small. The reason the Author has selected the four or five story building as the "cut off" point is that in most cases buildings above five stories are too high for package type equipment unless "floor by floor" units or "water source heat pump" equipment is used. In such cases even though the building may be much higher than five stories the methodology would be similar to that in a low or mid rise structure. These two types of systems will be discussed under the "Low & Mid Rise Building" paragraph.

Most so called high rise buildings are either office buildings or condominiums. These structures usually use chilled water or "Built Up" direct expansion (DX) systems to provide cooling. Heating for such systems is almost universally hot water.

Standard practice on high rise office buildings is to install the mechanical work in the "core." This is the center portion of the building which includes the elevators and restrooms. The air handling system for the building is located in the basement, roof penthouse or on building over 25 stories on an intermediate floor such as the 13th which is difficult to lease because of superstition.

During initial construction, the central plant and fan rooms are completed. Main supply duct risers extend through shafts from the fan rooms to all floors. At each floor, supply ducts are extended from the main duct in the shaft and form a loop around each floor, as is a hot water loop and mains extending from the boiler room.

Often the same duct shaft is used for return air. The shafts must be fire rated and all duct and return opening penetrations of the shaft wall to each floor must be equipped with an approved fire damper. Elevator lobbies, corridors and public areas are completed at this time by the contractor providing the "Core" work.

The Tenant areas of each floor are completed only after the tenant space is leased. The completion of these areas is referred to as "Tenant Improvement" or (TI.) Some HVAC contractors specialize on TI work in the larger metropolitan areas and leave the core work to the larger contractors. This work consists of installing connection to the main supply air "loops", terminal boxes, zone controls and related ductwork and air distribution. Where hot water coils are supplied in the terminal boxes, the TI contractor connects hot water piping from the main hot water loops to the coils within the terminal boxes.

If the system is supplied with chilled water, the chillers can be on the same floor as the air handling equipment. On some buildings the air handling equipment can be on more than one floor. In such cases the central plant can be located in the basement, a separate building or in a roof penthouse.

Where air handling equipment is 80,000 cfm or larger the air handling portion generally consists of a bank of coils, filter bank, supply fan(s) and return fan(s). Most modern systems incorporate an economizer cycle consisting of outside air dampers, return dampers and exhaust dampers. In this manner, outside air can be used to cool the building, when ambient temperatures are low, without the need for refrigeration. On built up DX systems, the compressors, condensers and cooling towers are often on the same level as the air handling room.

Most modern air handling systems are based

on the Variable Air Volume (VAV) principal. In such cases heating coils, electric or hot water, are located within the terminal boxes for the exterior zones.

The supply and return fans are usually of the vane axial type, plug fan type or centrifugal DWDI type. Fans operating at over 6" of static pressure are not uncommon.

On VAV systems there must be a method of modulating the air flow to maintain the design duct pressure at the terminal boxes, usually 1" or less. In the case of some vane axial fans, the pitch of the blades is altered from a control signal that repositions a motorized actuator. Although not encountered very often in modern systems, centrifugal DWDI fans may be equipped with inlet vanes which are modulated from a control signal. Most modern system use an electronic variable speed drive (VSD) which changes the speed of the fan wheel to maintain the required static pressure. VSDs are also referred to as adjustable frequency drives (AFD) or variable frequency drives (VFD.) Both supply and return fans are so equipped.

b)Low and Mid Rise Buildings

Buildings than four stories or less are considered mid rise or low rise. The reason the Author has selected the four or five story building as the "cut off" point is that in most cases these buildings lend themselves to package type equipment or water source heat pump systems. Many building of this type also utilize chilled water, hot water systems. These types of buildings with package type systems may be better suited to the beginning contractor unless he has had experience in the more complex central plant systems or the water source heat pump systems.

Low rise and mid rise buildings include office buildings, retail stores and a variety of other type occupancies. With the exception of office buildings, the entire cooling and heating system is generally installed during the original construction. Office buildings with package type systems generally have the equipment installed and the main supply and return ducts extended into the space. The final Tenant Improvement (TI) work is performed after the space is leased.

The same is also true with the water source heat pump systems where the boiler, cooling tower and piping is installed with the core. During the TI work, the heat pumps are installed as well as the zone ductwork and controls.

If the system is supplied with chilled and / or hot water, the chillers and / or boiler can be on the roof or can be located in the basement, a separate building or in a roof penthouse. The same is true of the cooling tower and boiler used with a water source heat pump system.

In some cases the upper floor(s) may use self contained package type units located on the roof. The lower floors often use "split systems" where the condensing units are located on the roof and air handlers or fan coil unit are used in the actual zones. These may be installed above a suspended ceiling or in an equipment room.

Most buildings of this type often use dedicated equipment for each zone, Variable Air Volume (VAV) systems or some type of zoned damper systems. In the case of VAV systems heating coils, electric or hot water, are located within the terminal boxes for the exterior zones.

When designing a building to use dedicated

units, one must be careful to separate interior zones from exterior zones. To neglect such will create a problem of constant complaints over certain rooms being too cold and some rooms being to warm. The reason for this is that as the solar load on the building changes, more summer cooling will be required in the exterior zones exposed to this load whereas the interior zones are not effected by solar loads.

b)Marketing & Sales

Where do you find projects and how can you obtain a contract? An excellent question which will now be addressed. If an existing business is being purchased the first thing to do would be to contact existing customers and find out about any projects being planned.

Another excellent method would be to contact some of the local consulting mechanical engineers and learn what projects are in the design stage. Even though the consulting engineer may not be in a position to assist you in getting contracts you can find out the name of the owner and Architect on different projects. You can then contact the owner or Architect and hopefully find out what General Contractors will be quoting on the project.

The new contractor should subscribe to the local construction newspaper or service. You will then learn who the active General Contractors are and what type of projects they are bidding. When you learn the names of some of the General Contractors who seem to be successful, learn the name of their estimator. Stop by their offices and get acquainted. There is nothing any better that establishing a personal relationship. When several very close sub bids are received by the estimator he will probably go with the sub contractor he is acquainted with. They can tell you what projects they are bidding and they generally have a plan room where you can review the plans and specifications and make your "take off."

In looking for residential projects you can follow up on sites where grading is taking place however on a commercial project when you see grading taking place you are too late. Generally when the ground is broken all bids have been received and contracts issued.

Most cities of moderate size will have a "Plan Room." In some cases the plan room is connected with a company which publishes a construction newspaper. This plan room is available to their subscribers at no cost. There are some plan room which are available for use by the members of an association which might be advisable for you to join.

In plan rooms you will find plans and specifications on most major projects in your area. It will be well worth your time to locate the nearest plan room and get acquainted and find out the requirements. The afore mentioned newspapers will list the projects that are out for bid as well as the General Contractors who are soliciting sub bids.

Become acquainted with local Architects. A call to the business manager at the School District can give you the name of the District's Architect. The next suggested step is to contact the Architect's office and learn the name of the Project Architect. This is the individual who is in charge of the particular school which you are interested in bidding. You can learn from the Project Architect when the job will be coming out to bid.

Bid openings on all "Public Works" are open to the public. It might be to your advantage to go to these bid openings and meet some of the General Contractor representatives. Generally those you might meet at bid openings may be a trusted secretary or other clerical person. Don't ignore these individuals (generally women.). They are a great source of information and are often the "gate keeper" to the decision maker. Treat these women with <u>great respect</u> and learn to call them by name. Never Ignore them because you might think they are a minor player. They can do you a world of good, like possibly getting you preferential treatment when it comes time to collect your payment. An occasional box of candy for the receptionist or the secretary, also a box of donuts for the office crew never hurts.

Marketing & Sales

Where do you find projects and how can you obtain a contract? If an existing business is being purchased the first thing to do would be to contact existing customers and find out about any projects being planned. Many buildings of this type are done on a Design / Build (D/B) and if approaching this type of market, the Contractor needs good engineering skills available. This can be done " in house" or an association with a local consulting mechanical engineer may be a good approach. You can also contact the owner or Architect and hopefully find out what General Contractors will be quoting on the project.

The new contractor should subscribe to the local construction newspaper or service. You will then learn who the active General Contractors are and what type of projects they are bidding.

When you learn the names of some of the General Contractors who seem to be successful, learn the name of their estimator. Stop by their offices and get acquainted. There is nothing any better that establishing a personal relationship. When several very close sub bids are received by the estimator he will probably go with the sub contractor he is acquainted with. They can tell you what projects they are bidding and they generally have a plan room where you can review the plans and specifications and make your "take off."

In looking for residential projects you can follow up on sites where grading is taking place. On commercial and institutional project when you see grading taking place you are too late. Generally when the ground is broken all bids have been received and contracts issued.

Most cities of moderate size will have a "Plan Room." In some cases the plan room is connected with a company which publishes a construction newspaper. This plan room is available to their subscribers at no cost. There are some plan rooms which are available for use by the members of an association which might be advisable for you to join. In such plan rooms you will find plans and specifications on most major projects in your area. It will be well worth your time to locate the nearest plan room and get acquainted and find out the requirements. The afore mentioned newspapers will list the projects that are out for bid as well as the General Contractors who are soliciting sub bids.

Bid openings on all "Public Works" are open to the public however smaller projects are generally private and public bid openings are not used. On D/B work the Contractor needs a good salesman who is capable of negotiating a contract. As is the case with larger projects become acquainted with the clerical help in the General Contractors office. Again, it is important to treat the clerical staff with great respect and learn to call them by name. Never Ignore them because you might think they are a minor player. They can do you a world of good - like possibly getting you preferential treatment when it comes time to collect your payment.

Chapter 2

HOW MUCH MONEY WILL I NEED?

An excellent question. Unfortunately the cause of the failure of most new businesses is not a lack of technology but is a lack of available capital to carry one through the problem jobs. Problems can consist of a poor paying general contractor or owner or a job will just "go sour" for a number of reasons. This is as good a time as any to define what makes a job "go sour" and the most common reasons are listed as follows:

1) Under estimating job costs.

2) Poor collections creating cash flow problems.

3) Credit problems with suppliers.

4) Labor disputes

5) Weather conditions which effect labor and job access difficulties.

6) Disputes with other subcontractors, the general contractor, Engineer or Owner.

7) Late delivery of equipment and materials.

Hopefully this list hasn't discouraged you. Remember the earlier comments about the employees making the boss rich? If you aren't discouraged at this point you probably are one of those individuals who has an excellent chance of becoming a successful contractor.

Now let's talk about the required capital. This is where "Murphy's Law" rules. You know the saying "Whatever can go wrong will." You can change this a little to say "Whatever the amount you calculate, it won't be enough."

There is a substantial difference between acquiring an operating business that has an established cash flow and starting a business "from scratch." In starting a new business you must consider that you have to bid some work, make the installation, submit your bill and then wait until the invoice is paid. Before you start your business, you need to consider what the start up costs will be.

Let's consider that you are going to start into the residential market and you will be performing the work yourself. Unless someone in your family has a job and can "bank roll" your business consider the following costs:

- A pick up truck or van
- Insurance on your truck or van
- Office and shop rental
- Office equipment and furniture
- Liability Insurance
- Business License
- Contractor's License
- Basic hand and power tools
- Minimum shop equipment (see chapter 3)
- Cash reserve until you collect your money from jobs (allow 90 days minimum)

Chapter 3

WHAT TYPE OF EQUIPMENT WILL I NEED?

The answer to this question depends on the type of business and the size of the operation. For the purpose of this chapter let's assume that the business is a new start up and there is not unlimited capital available from Uncle Murray's trust fund.

SERVICE BUSINESS

Other than the requisite office equipment you should plan on the following equipment for each service technician as a minimum:

1) Pick up truck with utility body or a van equipped with compartments for tools, parts and supplies.
2) Complete set of hand tools consisting of socket set, screw drivers, adjustable wrenches, combination box/end wrenches, pry bars and refrigeration specialty tools.
3) A set of refrigeration gages
4) Voltmeter, ammeter, ohmmeter, infra red thermometer
5) Superheat thermometers
6) Vacuum pump
7) Refrigerant reclaim unit
8) Electronic vacuum gage
9) Electronic leak detector or UV light detector
10) 100 feet 3 wire extension cord.
11) Electric impact wrench with sockets
12) Battery operated 3/8" drill with charging unit.

13) 5' step ladder
14) Hand truck
15) 14' extension ladder
16) Hi Jack or similar lifting device
17) Set of portable oxy / acetylene tanks with gages, torch and hoses.
18) Nitrogen tank and pressure reducing gage for pressure testing.

RESIDENTIAL (EXISTING HOMES)

This is a field where you will be doing it all. Sheet metal, refrigeration, carpentry, electrical, concrete, drywall repairs and painting. In addition to the equipment listed for a service technician, you will need the following for each truck as well as for your shop.

TRUCK EQUIPMENT
1) Electrical reciprocating saw
2) Electrical circular saw
3) Carpentry hand tools
4) Electric vibrating palm sander
5) Sheet metal hand tools
6) Drop clothes
7) Painting tools
8) Drywall tools

SHOP EQUIPMENT
1) 8' x 18 gage hand brake
2) Pittsburgh or button lock machine
3) 4' x 12' long work bench (lay out)
4) 3' x 20 gage slip rolls
5) Bench grinder

6) 15" drill press
7) 180 amp arc welder
8) Spot welder

RESIDENTIAL (NEW CONSTRUCTION)

If your main thrust will be new residential construction, and not the service business, you should plan on the following equipment as a minimum.

<u>PER TRUCK</u>
1) Pick up truck or 1 ton flat bed truck with stake bed. A lockable weather tight tool box should be permanently mounted in the bed to secure power tools.
2) Complete set of sheet metal tools.
3) Hand saw and carpenter's hammer
4) Electric or gasoline chain saw.
5) Shovel
6) A set of refrigeration gages (for start up)
7) Voltmeter, ammeter, ohmmeter, infra red thermometer (for start up)
8) Vacuum pump (for start up and service during warranty period)
9) Electronic vacuum gage
10) Electronic leak detector
11) 100 feet 3 wire extension cord and construction type adapters
12) Battery operated 3/8" drill with charging unit.
13) 5' step ladder
14) Hand truck
15) 14' extension ladder
16) Nitrogen tank and pressure reducing gage for pressure testing.

<u>SHOP EQUIPMENT</u>

1) 8' x 18 gage hand brake
2) Pittsburgh or button lock machine
3) 4' x 12' long work bench (lay out)
4) Bench grinder
5) 15" drill press
6) 180 amp arc welder
7) Spot welder

LIGHT COMMERCIAL & SCHOOLS (NEW OR EXISTING BLDGS)

If your main thrust will be light commercial construction, and not the service business, you should plan on the following equipment as a minimum.

PER TRUCK
1) Pick up truck or 1 1/2 ton flat bed truck with stake bed. A lockable weather tight tool box should be permanently mounted in the bed to secure power tools.
2) "Gang Box" (rolling tool box for securing tools on site)
3) "Come Along" cable hoist
4) Complete set of sheet metal tools.
5) Caulking gun
6) Set of refrigeration gages (for start up)
7) Voltmeter, ammeter, ohmmeter, infra red thermometer (for start up)
8) Vacuum pump (for start up and service during warranty period)
9) Electronic vacuum gage
10) Electronic leak detector
11) 200 feet 3 wire extension cord and construction type adapters
12) Battery operated 3/8" drill with charging unit.
13) 8' step ladder

14) 10' step ladder
15) "Hi Jack" or similar lifting device.
16) Hand truck
17) 20' extension ladder
18) Nitrogen tank and pressure reducing gage for pressure testing.

SHOP EQUIPMENT
1) 8' x 18 gage hand brake
2) Pittsburgh or button lock machine
3) 10' x 16 gage power squaring shear
4) Duct notching machine
5) 4' x 12' long work bench (lay out)
6) 4' x 12' long bench (duct lining)
7) Bench grinder
8) 15" drill press
9) 180 amp arc welder
10) Spot welder

HEAVY COMMERCIAL (NEW OR EXISTING CONSTRUCTION)

If your main thrust will be doing this type of work be aware you are entering the "Big Leagues" and should be very experienced in this type of construction. The shop equipment will be dependant on the size and scope of work you are undertaking. For the purpose of this book we will assume that you will be using a sub contractor for your piping.

PER TRUCK

1) Pick up truck or 1 1/2 ton flat bed truck with stake bed. A lockable weather tight tool box should be permanently mounted in the bed to secure power tools.
2) "Gang Box" (rolling tool box for securing tools on site)
3) "Come Along" cable hoist
4) Complete set of sheet metal tools.
5) Caulking gun
6) Set of refrigeration gages (for start up)
7) Voltmeter, ammeter, ohmmeter, infra red thermometer (for start up)
8) Vacuum pump (for start up and service during warranty period)
9) Electronic vacuum gage
10) Electronic leak detector
11) 200 feet 3 wire extension cord and construction type adapters
12) Battery operated 3/8" drill with charging unit.
13) 8' step ladder
14) 10' step ladder
15) "Hi Jack" or similar lifting device.
16) Hand truck
17) 20' extension ladder
18) Nitrogen tank and pressure reducing gage for pressure testing.

SHOP EQUIPMENT

1) 8' x 18 gage hand brake
2) 4' x 18 gage hand brake
3) Pittsburgh or button lock machine
4) 10' x 16 gage power squaring shear
5) 10' x 16 gage press brake
6) Duct notching machine
7) Insulation pin spotter

8) Computer operated laser cutting line
9) Slip and Drive forming machine
10) Pocket lock (Government Lock) forming machine
11) 4' x 12' long work bench (lay out)
12) 4' x 12' long bench (duct lining)
13) Bench grinder
14) 15" drill press
15) 180 amp arc welder
16) Spot welder
17) 2 ton stake bed truck for delivery of ductwork to job site

The above listing of equipment for the various types of work is the writer's opinion gleaned from experience. As you become more familiar with your line of contracting there will be additional equipment and tools you may wish to add to speed up your work These lists are to allow you to put a dollar figure by each item to see what the capital required will be.

Chapter 4

DIFFERENT SYSTEMS TO BE ENCOUNTERED

METHODS OF DISTRIBUTING HEATING & COOLING

Cooling may be distributed in two different manners.

1) Direct Expansion (DX) systems extract the heat directly by passing of air through a finned surface evaporator coil. Refrigerant is boiled directly within these tubes to absorb the heat and eventually rejecting said heat at the condenser.

2) Chilled water system make use of finned surface coils within the air stream. Chilled water is generated at a remote point, from a chiller, and pumped to multiple cooling coils throughout the facility. The heat from the air stream is extracted and transferred to the cooling coils by chilled water. This heat is eventually removed by the evaporator in the chiller and rejected through the chiller's condenser.

Dedicated chilled water piping may be used in conjunction with a piping system carrying hot water. Such is referred to as a "4 Pipe System.' There is a combination system where the same piping is used for either chilled water or hot water depending on the season. Such is referred to as a "2 Pipe System."

Heat is distributed in a similar manner.

1) A furnace can be fired by oil, gas, coal, wood or any other combustible fuel. The products of combustion from the fire box pass through a heat exchanger located in the air stream. The heat is transferred through the walls of the heat exchanger and is absorbed by the air passing through thus delivering heat to the space.

2) Gas fired or electric radiant heaters do not utilize an air stream to move the heat. Heat is from direct infra red radiation contacting the body. The air surrounding the body is not heated. This is similar to standing in the sun in weather below freezing and feeling warm. Other solid objects absorb the infrared radiation also and well in turn re-radiate the heat until the air is warmed. This is a secondary form of heating as the direct radiation is primary.

3) Electric heating elements can be located in the air stream to provide heating.

4) Another all electric system is the heat pump. This is some times referred to (although inaccurately) as a reverse cycle system. More about heat pumps in later paragraphs.

5) Finned coils located within the air stream can be supplied with hot water, from a central boiler plant, or with steam from a steam boiler. In either case the cooler air stream passing through the heating coil absorbs heat from the hot water or steam.

1). SYSTEMS WITH ECONOMIZERS

A. Air Side Economizers

An "Air Side" economizer cycle may be incorporated within DX package equipment. Unit ventilators or any type of air handling equipment. This arrangement allows the use of outside air for cooling purposes when the climatic conditions are right and uses three dampers to provide this feature. An outside air damper which allows outside air to enter the supply fan compartment, a return damper to prevent the return air from entering the supply fan compartment and an exhaust damper which exhausts the air from the conditioned space being returned to the equipment through the return air ductwork.

There are two basic types of controls to accomplish this feature, "Dry Bulb Economizer Controls" and "Enthalpy Economizer Controls." The dry bulb control system places the dampers into the economizer mode and shuts down the refrigerated cooling when outside temperature is below a certain set point, usually 60 degrees F. When the wall thermostat in the conditioned space reaches it's set point, the economizer dampers return to their original position.

The enthalpy control places the dampers into the economizer mode and shuts off the refrigerated cooling when a combination of outside dry bulb and wet bulb conditions reach a certain set point. This arrangement has the advantage over the dry bulb system inasmuch as if the outside air is below 60 degrees and the relative humidity is high, the equipment will remain in the refrigerated mode and not add unwanted moisture to the conditioned space.

Either of these control systems may be two

position (in economizer mode or refrigerated mode) or modulating where the outside air is modulated to maintain a given supply air temperature. The room thermostat can then return the dampers to their original position when the room thermostat reaches a predetermined set point.

In systems with more than a minimum of return air duct, a return fan or power exhaust fan is incorporated. This prevents the pressure in the condition space from increasing mover a given set point. On small systems, less than 20 tons, a separate roof exhauster with a damper is generally incorporated. In such cases a duct connecting the return duct and the exhaust fan is used. Upscale small equipment (3 tons through 20 tons) often have a power exhaust fan built into the equipment. Larger systems (above 20 tons) almost always have a return fan air fan or power exhaust fan incorporated into the unit.

B) Water Side Economizers

There are systems using a "Water Side Economizer." These operate very similar to the "air side" system using "Dry Bulb Controls." Such systems use a cooling tower, pump and piping to circulate water through a dedicated coil located within the return air side of the system. This coil can be built into the equipment or placed into the return air ductwork.

A modulating control valve is used to maintain the supply air at a predetermined temperature. This type of arrangement is used where it is not practical to use outside air ductwork (other than that required for minimum ventilation) and exhaust air ductwork.

2). SYSTEMS WITH HEAT RECOVERY

Heat recovery systems are energy saving devices which can be applied to both cooling and heating systems. Such systems reclaim energy (heat or cool) from exhaust streams and transfer the savings to the air stream of cooling and/or heating system. Some systems transfer just sensible heat and some transfer total heat which includes the adding or removing of moisture.

Some of the more commonly encountered devices are as follows:

A) Run Around System

A "Run Around" system, which uses a finned heat recovery coil in the exhaust stream and another similar coil in the outside air stream or the return air stream from the occupied space, is used for transferring sensible thermal energy.

The fluid transferring such energy is either water or a water / glycol mixture which is pumped from one coil to the other in a continuous loop. The heat recovery coil can also transfer heat from the flue or chimney of a combustion device such as a boiler or furnace.

B) Air to Air

An all air device such as an "Air to Air" heat exchanger will transfer energy from one air stream to another. This is a static device with no moving parts in either a counter flow or cross flow configuration.

The heat transfer surface can be aluminum. copper,

stainless steel and, depending on the application, plastic or special treated paper (cellulose.)

C) Heat Pipe

Another sensible heat transfer devise is the "Heat Pipe." Such device consists of a series of tubes and fins which resemble a conventional hot water or chilled water coil. One portion of the coil is located in the warm air stream and the other in the cool air stream.

The major difference is that there are no return bends as each tube is an independent heat transfer unit. Within each tube is contained a high boiling temperature refrigerant such as R11 and a wicking material which absorbs the refrigerant. The tubes are slightly pitched.

The section of the tube located in the air stream being cooled functions as the evaporator and the section located within the air stream being heated functions as the condenser. The liquid refrigerant is boiled thus cooling the surface of the coil which absorbs heat from the air stream being cooled. The refrigerant vapor raises to the upper level of the pitched tube where it is condensed and returns to the evaporator section where the process is repeated.

Depending on which section is pitched upward, the heat pipe can be used to recover either cooling of heating. Some such devices have a damper actuator and a tilting mechanisms where the unit can be pitched for use as either a heating or cooling energy recovery device.

D) Heat Wheel

The "Heat Wheel" is another common air to air device

which has a rotating media comprised of corrugated aluminum or steel material. One portion of the wheel is exposed to the warm air stream and another portion of the wheel is exposed to the cool air stream.

Depending on the type of application the device is designed for, a small portion of the wheel (less than 5%) is used for purge. The purpose of such purge is to minimize and air bourne contaminants in the exhaust stream from entering the conditioned air stream.

The heat transfer media can be coated with a desiccant material which absorbs moisture from one stream and transfers the moisture to the other stream. Such is described as "Total Heat" recovery as opposed to "Sensible Heat" recovery only.

3). DIRECT EVAPORATIVE COOLING SYSTEMS

Direct evaporative cooling actually does not transfer any heat. It simply converts sensible heat into latent heat thus lowering the dry bulb temperature of the air. The efficiency of such systems is based on an air temperature drop as a percent of the wet bulb depression (Dry bulb - wet bulb temperature.) An 80% efficient unit would provide the following supply air when using the following examples:

Ambient Dry Bulb	100 Degrees F
Ambient Wet Bulb	- 65 Degrees F
Wet Bulb Depression	35 Degrees F
Efficiency of Unit	80%
Sensible Cooling Effect	28 Degrees F

Ambient Dry Bulb	100 Degrees F
Cooling Effect	-28 Degrees F
Supply Dry Bulb	72 Degrees F

As is evident from this example, assuming sufficient air changes are used, the space could be cooled satisfactory. It should be remembered that no actually cooling is taking place as the evaporated moisture is added to the conditioned air stream and the air is almost 100% saturated and the relative humidity of the air would be approximately 75%. If however the ambient air conditions are as shown in the following example, the results using the same equipment would also be far different and produce a relative humidity of approximately 85%.

Ambient Dry Bulb	95 Degrees F
Ambient Wet Bulb	-78 Degrees F
Wet Bulb Depression	17 Degrees F
Efficiency of Unit	80%
Sensible Cooling Effect	13.6 Degrees F
Ambient Dry Bulb	95 Degrees F
Cooling Effect	-13.6 Degrees F
Supply Dry Bulb	81.4 Degrees F

A) Pad Type (Swamp Cooler)

The most commonly encountered direct evaporative cooler (often referred to as a "swamp cooler") is one using evaporative media made of wood fibers, usually aspen wood. A fan draws outside air across the media pads which are saturated with water which has been pumped from a reservoir to distribution troughs located above the media pads.

This type of cooler is available in both horizontal and down shot configurations. This type of unit is generally limited to residential and warehouse type applications. Depending on the particular unit the efficiencies are between 65% and 80% efficient. The wood fiber pads need replacement annually. Water is continually bled off to minimize mineral depositing on the wood fibers thus blocking the air flow.

B) Honeycomb Type (CelDek)

CelDek is a trade name of the Munters Corporation. And as they were the originator of this type of high efficiency media, the trade name has become synonymous with this type of media.

This type of direct evaporative cooler is found on most schools, commercial and industrial applications and uses a "honeycomb" evaporative media from between 4" and 24" thick. The efficiency of this media runs from 75% to 90% + depending on face velocity and thickness.

This material is made from paper (cellulose) impregnated with epoxy resin. A material made from glass fibers is also available. Air velocities should be kept at 550 feet per minute or lower to prevent water droplets from carrying over and entering the duct system.

The water is distributed in a somewhat similar manner as with the wood fiber pad type media. The life expectancy of such material is 5 to 10 years depending on the annual hours of use. Water bled off is required to minimize mineral depositing within the air passageways.

C) Rotary Coolers

Rotary coolers are those which use a copper mesh media wheel and were used extensively by the U.S. Military. These wheels rotated slowly through a reservoir of water thus exposing the wetted media to the air stream. The efficiencies are between 70% and 80% depending on the thickness and speed of the wheel as well as the air velocity over the media.

This type of device is rarely encountered today due to the high efficiency and lower cost of the honeycomb media type evaporative coolers.

D) Air Washers

Air washers are sometimes used in large air flow quantity applications in industrial buildings. With such installations water is sprayed against a spray deck media of some type or a series of metal louver type blades. These systems were popular in the 1950s and 1960s but are seldom seen today.

Moisture eliminators are located down stream from the spray deck to remove water droplets from the air stream. The efficiency of such systems are between 75% and 90% depending on face velocity and depth of the spray deck.

The advent of the honeycomb type media has almost eliminated the use of this type of system.

E) Spray Nozzles

Atomizing spray nozzles, sometimes referred to as "Fog Nozzles", often using compressed air or water at super sonic velocities are sometimes used in air handling systems or direct into the space in some industrial applications. The use of compressed air or super sonic water velocities from the nozzles break the water into a fine mist.

a) Some manufacturers use laser bored ruby orifices to prevent erosion and are sometimes referred to as fog nozzles.

4). INDIRECT EVAPORATIVE COOLING SYSTEMS

Indirect evaporative cooling provides non refrigerated air cooling without adding moisture to the air stream. As opposed to direct evaporative cooing which provides no heat transfer, the indirect evaporative cooling system actually removes heat from the air stream and rejects the heat to the outside of the building, a true heat transfer.

There are two basic methods of indirect evaporative cooling which are described as follows:

A) Cooling Tower Method

The "cooling tower" method evaporates water in either a separate cooling tower or a similar device is incorporated within the evaporative cooling unit.

By evaporation the water is cooled to within 2 to 3 degrees of the outdoor wet bulb temperature. This cooled water is then pumped to a finned surface coil located within the air stream being conditioned.

A properly sized coil can provide discharge air within 4 degrees of the water temperature entering the coil.

Example:
Outside dry bulb	95 Degrees F
Outside wet bulb	72 Degrees F
Approach	3 Degrees
Water Entering Coil	75 Degrees
Air Entering Coil	95 Degrees
Air Leaving Coil	79 Degrees

This system can utilize heat recovery by using exhaust air from the building to evaporate the water within the cooling tower. In such cases the building exhaust air wet bulb temperature may be much lower than the outside dry bulb temperature and therefore create a lower water temperature to the cooling coil.

Example:
Outside dry bulb   95 Degrees F
Exhaust air wet bulb   68 Degrees F
Approach   3 Degrees
Water Entering Coil   71 Degrees
Air Entering Coil   95 Degrees
Air Leaving Coil   75 Degrees

B) Heat Exchanger Method

Another method, which is generally a little more efficient provides a heat exchanger where water is allowed to evaporate on the surface of the heat exchanger lowering the surface temperature to the ambient wet bulb temperature.

The opposite side of this exchanger is exposed to the air stream being conditioned. Depending on the efficiency of the unit, the discharge air can be cooled within 4 degrees of the heat exchanger surface temperature.

Example:
Outside dry bulb   95 Degrees F
Exhaust air wet bulb   68 Degrees F
Approach to HX Temp   4 Degrees
Air Entering Coil   95 Degrees
Air Leaving Coil   72 Degrees

5) INDIRECT / DIRECT EVAPORATIVE COOLING SYSTEMS

The indirect / direct evaporative method, as it's name implies, combines both the direct and indirect methods. Depending on the climatic conditions, the performance will approach that of a refrigerated cooling system. This system has a higher cost than a package air conditioning unit but the energy savings can provide a short payback period. EERs can be as high as 50.

Example:
Ambient air entering the indirect stage	95 fdb/ 68 fwb
Air leaving indirect and entering direct section	72 fwb/ 60 fwb
Air leaving 90% efficient direct section	60.2 fwb/ 60 fwb

6) MAKE UP AIR SYSTEMS

These systems are normally found in the kitchen area or school gyms. In the kitchen areas these units bring in 100% outside air to replace the air exhausted from the range hood and other exhaust systems in the kitchen. These systems consist of an air handling unit containing a direct evaporative cooler section and some method of providing heat. The cooling section can be of the direct evaporative type or in extremely humid climates be of the indirect / direct evaporative type or actually a chilled water coil.

The heating portion is sized to temper the air being

introduced into the space as generally there is enough internal heat to offset building heat loss. This heat may be through gas fired furnaces or through steam or hot water coils. When indirect gas fired furnaces (those with heat exchangers) are used stainless steel is generally specified for the heat exchanger element. The burners are either modulating or set up for "low fire / high fire."

7) WATER SOURCE HEAT PUMP SYSTEM

This type of system is a substitute for a 4 pipe chilled/hot water systems. Individual small heat pumps are dedicated to each zone. They are different in that they have no air cooled condenser and fan, traditionally used with air source heat pumps, and instead use a refrigerant to water heat exchanger. These heat pumps are available in a vertical or a horizontal configuration.

A closed circuit cooling tower or fluid cooler is provided and water is circulated through all heat pumps where compressors are operating. When the heat pumps are in the cooling mode, the heat is transferred into the water which is then rejected through the cooling tower.

When some units require heating, the warm water from the units that are in the cooling mode become the heat source for those in the heating mode. Through the period of the year where both cooling and heating are required this system is highly efficient. When the majority of the heat pumps are in the heating mode, the water circulated to the cooling tower becomes quite cold. To prevent the piping from sweating (insulating of these lines is rarely done) an auxiliary boiler is fired to maintain the water temperature above the surrounding dew point. The use of this boiler is minimal so the operating cost for this auxiliary heat is relatively low. The auxiliary boiler may use what ever type of fuel that is economically available.

8) FAN COIL SYSTEM

This type of system uses chilled water for cooling and either hot water or electric resistance heaters for heating. These units are generally equipped with small direct driven fans and are not capable of more than a very minimum amount of ductwork.

There is no provision for an economizer cycle just a minimum of outside air. They are equipped with a chilled water coil and insulated drain pan with a secondary pan specified. This secondary pan may be by the manufacturer but more than often must be provided by the installing contractor.

Fan coil units are available with decorative cabinets, for exposed application or with no cabinet for installation where they are not visible. They may be of a horizontal or vertical configuration. Such units are generally in sizes between 300 cfm and 1,200 cfm. Some manufacturers produce slightly larger units. The controls may be suppled by the unit manufacturer or be field supplied and installed by the temperature control contractor.

9) CHILLED WATER UNIT VENTILATOR SYSTEMS

Unit ventilators are quire similar to fan coil units. The major difference is that they are equipped with economizer dampers capable of 100% outside air. The most commonly encountered installation is at the outside wall of the building with an outside air louver and sleeve provided. There are no exhaust fans built into these units so a separate relief damper and louver must be used.

There are units designed specifically for classroom applications in a vertical configuration with adjoining matching book shelves. This type of unit is located under a window with the book shelves extending the full length of the wall. In some instances two unit ventilators are used. There are many option and types of these shelves. Some have open shelving and some have doors and are available as supply cabinets.

The largest of this type generally encountered is 2,000 cfm equaling 5 tons. The controls may be supplied by the unit manufacturer or be field supplied and installed by the temperature control contractor.

10) DX PACKAGE UNIT VENTILATOR SYSTEM

This type of unit ventilator is similar to the type that uses chilled water for cooling. The only difference is that in lieu of chilled water a refrigerated condensing unit is built into the unit. Heating may be through a hot water coil served from a central boiler plant, electric resistance heat or the unit may be ordered as a heat pump. A louver for condenser air is provide along with the outside air louver.

HEATING AND VENTILATING UNITS

These units operate very similar to a unit ventilator except there is no provision for cooling. They are available in larger sizes and are commonly encountered in gyms and locker rooms when no cooling is required.

11) 2 DECK MULTIZONE CHILLED WATER AIR HANDLING SYSTEM

Such system are used where a single air handing unit is used to supply multiple temperature control zones and are available in horizontal, up flow or down flow configurations. These units are available in sizes from 6,000 cfm through 100,000 cfm and are available with gas, oil, hot water, or steam heat as the heating medium and must be of the "blow through" configuration. Cooling is provide from a central chilled water plant.

The units can be indoor type or are available for roof installation. They can be supplied with return fans, power exhaust fans and with economizer cycles and with heat recovery systems. A separate supply duct connect from the zone head of the unit to a specific zone. A common return duct is used to return air from all of the zones to the unit.

As the zone damper actuators modulate from a signal from the room thermostats, mixing occurs between the cooling air and the heating section. The mixing of cool air and warm air controls the space temperature. In some areas of the country this type of unit may no longer be used because of the energy consumed in mixing the two air streams for temperature control.

12) 2 DECK MULTIZONE DX PACKAGE SYSTEMS

Such system are used where a single package air conditioning and heating unit is used to supply multiple temperature control zones and are available in horizontal, up flow or down flow configurations. These units are available in sizes from 15 tons through 350 tons and are available with gas, oil, hot water, or steam heat as the heating medium and must be of the "blow through" configuration. This means that the heating and cooling sections must be downstream of the supply air fan. Cooling is provided from a condensing section with either reciprocating compressors or screw compressors.

Condensing may be through the use of a built in air cooled condenser coil, a built in evaporative condenser or water cooled condenser which requires a separate cooling tower and pump. Multiple compressors with capacity control are required. Hot gas bypass is required on the lead compressor so the unit will remain operational when very little air is passing through the evaporator coil.

The units are available for roof or grade level installation. They can be supplied with return fans, power exhaust fans and with economizer cycles and with heat recovery systems. A separate supply duct connect from the zone head of the unit to a specific zone. A common return duct is used to return air from all of the zones to the unit.

As the zone damper actuators modulate from a signal from the room thermostats, mixing occurs between the cooling air and the heating section. The mixing of cool air and warm air controls the space temperature. In some areas of the country this type of unit may no longer be used because of the energy consumed in mixing the two air streams for temperature control.

13) 3 DECK MULTIZONE AIR HANDLING SYSTEM

Such systems are applied where a single air handing unit is used to supply multiple temperature control zones and are similar to the more commonly encountered 2 deck units. The only difference being that there is a third damper section which handles air which bypass both the cooling and heating sections. The unit must also be of the "blow through" configuration.

As the zone damper actuators modulate from a signal from the room thermostats mixing occurs between the bypass air and the cooling air or between the bypass air and the heating section. There is no mixing of cool air and warm air. In some areas of the country this is the only type of multizone system that can be installed as the simultaneous operation of cooling and heating within a given space is prohibited by code.

A separate supply duct connect from the zone head of the unit to a specific zone. A common return duct is used to return air from all of the zones to the unit.

These units are available in sizes from 6,000 cfm through 100,000 cfm and are available with gas, oil, hot water, or steam heat as the heating medium. The units are available for roof or interior installation. They can be supplied with return fans, power exhaust fans and with economizer cycles and with heat recovery systems.

14) 3 DECK MULTIZONE DX PACKAGE SYSTEMS

Such system are used where a single air handing unit is used to supply multiple temperature control zones and are similar to the more commonly encountered 2 deck units. The only difference being that there is a third damper section which handles air which bypass both the cooling and heating sections. The units must be of the "blow through" configuration. This means that the heating and cooling sections must be downstream of the supply air fan.

As the zone damper actuators modulate from a signal from the room thermostats mixing occurs between the bypass air and the cooling air or between the bypass air and the heating section. There is no mixing of cool air and warm air. In some areas of the country this is the only type of multizone system that can be installed as the simultaneous operation of cooling and heating within a given space is prohibited by code.

A separate supply duct connect from the zone head of the unit to a specific zone. A common return duct is used to return air from all of the zones to the unit.

These units are available in sizes from 15 tons cfm through 350 tons and are available with gas, oil, hot water, or steam heat as the heating medium. The units are available for roof or grade installation. They can be supplied with return fans, power exhaust fans and with economizer cycles and with heat recovery systems.

Cooling is provide from a condensing section with either reciprocating compressors or screw compressors. Condensing may be through the use of a built in air cooled condenser coil, a built in evaporative condenser or water cooled condenser which requires a separate cooling tower and pump. Multiple compressors with capacity control are required. Hot gas bypass is required on the lead compressor so the unit will remain operational when very little air is passing through the evaporator coil.

15) VAV SYSTEMS AIR HANDLING SYSTEM

Such system are used where a single air handing unit is used to supply multiple temperature control zones using VAV terminal boxes in the ductwork. The air handlers may be of the "blow through" or "draw through" configuration.

As the zone damper actuators modulate within the VAV boxes, from a signal from the room thermostats, the change in duct static pressure is sensed and the air flow from the unit is reduced to the design static pressure of the terminal box, usually between .5" and 1.0". The air quantity is reduced by either fan inlet vanes or fan inlet cone which are controlled from a damper actuator. The method preferred by the Author is using an electronic Variable Frequency Drive (VFD) which changes the speed of the motor thus changing the air flow of the fan. As well as being the most energy efficient it is cost comparable when one considers the mechanical methods which require the use of a damper or cone, the actuator, the controller, magnetic motor starter and a disconnect switch, all of which are replaced by the VFD.

These units are available in sizes from 1,200 cfm through 100,000 cfm and are available with any combination of filters. Heating for the VAV system is accomplished within the VAV terminal box which contains a heating coil, either hot water or electric resistance. The units are available for roof, grade or interior installation. They can be supplied with return fans, power exhaust fans and with economizer cycles and with heat recovery systems.

16) SMALL PACKAGE SYSTEMS

These air cooled condensing package units are available in sizes from 2 tons through 15 tons and are available with natural gas, propane, oil, hot water, steam or as heat pumps for the heating mode. The units are available for roof or grade installation. Some units can be supplied with, power exhaust fans and with economizer cycles and with heat recovery systems. These systems are probably the most commonly encountered for residential, school classroom, and light commercial applications. These units may be available for either horizontal duct connections or "down shot" duct connections.

If roof mounted they are placed on a field built platform or curb or they may be furnished with a factory built curb. The factory built curb may be of the "knocked down" type or factory assembled and 8" to 14" in height. The roof platform is generally used when the roof is pitched more than 1/4" per foot. On these platforms a sheet metal cover may be specified or if the unit is a "down shot" type a low (6" to 8" high) roof curb should be place on top of the platform.

On those roofs with a pitch of 1/4" per foot or less, the curb should be shimmed so that the unit is dead level. Prefabricated roof curbs are generally available in heights of from 8" to 14". Units are rarely placed on grade with the exception of installations on existing buildings or metal buildings where roof mounting is not practical.

17) SMALL SPLIT SYSTEMS

Such systems consist of an air cooled condensing unit locate outside of the structure, normally on grade, and a furnace with an evaporator coil or air handler located at a remote location. Heating may be through the use of a furnace, hot water coil, steam coil, electric resistance heater or may be of the heat pump

The indoor section (furnace or air handler) may be located within a separate room or exposed within the conditioned space. The indoor unit may be placed on a raised platform or plenum to draw return air from a grille(s) just above the floor line. Supply air may be discharged from a decorative plenum when the unit is installed within the conditioned space or through concealed ductwork. The indoor unit is generally installed adjacent to the outside wall so that outside air for ventilation can be supplied through an exterior wall louver with manual damper.

Such units are not supplied with an economizer cycle but such may be supplied by a custom fabricator of fabricated by the installing contractor. Refrigeration liquid and suction lines must be used to connect the condensing unit to the evaporator coil. If there is more than 8 feet difference in the elevation between the air cooled condenser and the indoor evaporator coil a suction line trap should be provided. The suction line must be insulated to prevent condensation. A condensate line must be supplied to remove condensate from the drain pan to either a dry well or other method of disposal outside of the building.

18) GAS FIRED ABSORPTION CHILLER PLANTS

This type of system uses what is referred to as gas fired absorption refrigeration to produce chilled water. After the chilled water is produced it can be pumped through insulated piping and circulated through cooling coils located within a variety of air handling equipment.

The question is often asked," How can you get cold by adding heat?" The question is legitimate. If one remembers his high school physics, and possibly some of that boring chemistry course, it should be simple. The major components within the absorption cycle are very similar in function to those in the standard compression refrigeration cycle with possibly one exception.

A) HEAT SOURCE
In the case of dual effect absorption chillers, heat from natural gas, steam or hot water is the energy source.

B) REFRIGERANT
The refrigerant in the Lithium Bromide (Li Br) absorption systems is water. Because the refrigerant is water, the temperature of the chilled water used for transferring cooling between the chiller and the load is generally never lower than 43 or 44 degrees. The refrigerant (water) is generally 10 degrees lower than the circulating chilled water and at the above temperatures would be 33 to 34 degrees, close to freezing. There are small tonnage air cooled absorption units which use ammonia as the refrigerant and water as the absorbent but are not the subject of this paper.

C) ABSORBENT
The absorbent in the conventional higher

temperature absorption chillers is Lithium Bromide. Li Br is a non toxic salt that has the ability to absorb water (the refrigerant).

D) HIGH TEMPERATURE GENERATOR

The generator is the equivalent of the compressor in the conventional compression refrigeration cycle. It is a vessel where heat is applied which causes the mixture of refrigerant and absorbent to boil and provides the motive force for the process. The refrigerant vapor (steam) and absorbent droplets (Li Br and H2O) leave the generator and enter a separator.

E) LOW TEMPERATURE GENERATOR

Hot refrigerant vapor from the separator passes through a heat exchanger before entering a coil within the low temperature generator. This vapor (water) gives up heat to, and causes re-boiling of, the Li Br solution which has been partially concentrated within the high temperature generator during the initial boiling off of the refrigerant (water). The Refrigerant vapor from the re-boiling, as well as the refrigerant vapor within the coil, passes to the condenser.

F) CONDENSER

The vapors pass from the low temperature generator to the condenser where cooling water from a cooling tower circulates through the condenser removing the heat added at the generator. The refrigerant vapor is condensed into a liquid refrigerant where it passes through an orifice into the evaporator which operates under a vacuum. This performs the same function as the condenser in a conventional compression refrigeration system.

G) EVAPORATOR

The evaporator serves the same function as in a conventional compression refrigeration system inasmuch as it absorbs the heat from the process load. The expansion of the refrigerant (water) within a vacuum causes boiling at approximately 34 degrees and therefore absorbs heat from the chilled water coils located within the evaporator. This chilled water, at approximately 44 degrees, is used to satisfy the process load.

H) ABSORBER

The refrigerant (water) vapor will not condense under the vacuum required for evaporation. Some means of converting the vapor back into a liquid must take place before the refrigerant (water) is returned back to the generator to repeat the process.

This is where the absorption cycle differs from the conventional compression refrigeration cycle. The Li Br solution, which was concentrated within the low temperature generator when the refrigerant (water) was boiled off, passes from the low temperature generator through a heat exchanger to the absorber.

Cooling coils within the absorber absorb the heat load from the evaporator and, unavoidably, some of the residual heat from the concentrated solution, to be dissipated through the cooling tower. The concentrated Li Br solution absorbs the refrigerant (water) vapor from the evaporator. This Li Br / H2O solution is now diluted and is pumped through the heat exchanger, to recover heat from the concentrated solution, before returning to the high temperature generator to repeat the process.

When the refrigerant vapors are absorbed, a vacuum is created which allows the refrigerant from the condenser to expand into the evaporator causing the refrigerating effect.

I) CHILLER SIZES
Dual Effect Gas & Steam Fired Chillers are available

from several firms in sizes from 50 tons through more than 1500 tons. The Author believes that installations of less than 800 tons should use multiple smaller units because of redundancy and operating economies.

J) REMARKS

Dependent on the type of absorption units specified, the same equipment can also be used for heating. This is accomplished by passing the hot refrigerant (water) vapors directly from the high temperature generator to the evaporator where the chilled water coils be come heating coils. This is the function of the conventional gas fired "Chiller/Heater. This of course creates a "2 Pipe" system.

When simultaneous cooling and heating is required, such as in a "4 Pipe" system, a "Double Duct " or "Multizone System", the percentage of cooling and heating being provided can be determined by using the multiple modular concept. In this method the amount of chiller/heaters in the cooling or heating mode is determined by the actual load.

There is little argument that gas fired dual effect absorption chillers will operate more economically in most geographical areas, due to the differential in the cost between gas and electric energy. Because the capital cost of dual effect absorption equipment is greater than equivalent electric driven equipment, it is necessary to evaluate the substantial reduction in electrical service size, switch gear, cabling and the lack of need for an equipment room when using absorption chiller/heaters.

Many of the smaller sizes of absorption chillers, 100 tons and less, are designed for outdoor installations and require no equipment rooms as compared with larger size absorption chillers which do. The use of multiple small units can reduce pumping costs through the use of variable speed drives. Substantial savings can be realized by matching the chilled and cooling tower water flow rates with the quantity of chiller modules in operation.

Because the cost per ton for the smaller modular chillers is somewhat higher than that of the larger absorption machines, the economics must be carefully evaluated. Because separate boiler and chiller rooms are not required, these outdoor type units can show a substantial savings in the elimination of equipment rooms.

K) THERMAL ENERGY STORAGE PLANTS

a) GENERAL INFORMATION

TES systems are sometimes encountered in larger middle schools, high schools, college campuses, office buildings and hospitals. The basic theory of a TES system is to produce a cold storage mass during a time period where electrical rates are low and ambient temperature conditions are conducive to efficient refrigeration operation and then using this stored cooling during periods of high electrical rates and high operating conditions.

There are two distinct methods of storing cooling; Chilled Water Storage Systems and Phase Change Storage Systems. The first method makes use of sensible cooling only and the second method makes use of both sensible and latent cooling. The references to "brine" in this section denotes a liquid with a depressed freezing point such as a glycol or a chloride solution. A more correct term would be "secondary refrigerant" but the older "brine" term has been used to avoid confusion.

b) CHILLED WATER STORAGE

The Chilled Water Storage method would seem, at first glance, to be the simplest and most effective method. A standard water chiller can be used and the chilled water is produced at approximately 38 degrees and stored in an insulated tank. Because only sensible cooling is employed considerable volume is required.

Major considerations in such a system is the expense of the huge insulated tank, the unwanted blending of warm return water with the cold water within the tank as well as the hydraulic design problems of such a system. Blending can be minimized by using a chilled water temperature differential of from 20 to 30 degrees. This large temperature differential establishes a thermocline within the storage tank which separates the warm return water from the chilled water being stored.

To accomplish this large differential a small recirculating pump located at each building in a campus type application or is sometimes located at each cooling coil to allow the water temperature within the coil to arrive at a predetermined temperature before being allowed to return to the storage tank. A complex control system is required to accomplish this function. Coil selection is critical because of the relatively high water temperature differential.

Such a system often allows smaller chilled water piping and pumps. This not only reduces initial costs but saves on pumping costs as well as the economies of operating the chiller at the suction temperatures required to produce 38 degree water.

c) PHASE CHANGE SYSTEMS - GENERAL

Methods employing the phase change of the storage media either employ a eutectic salt or ice as the storage media. In this manner both sensible cooling as well as the latent heat of fusion is employed to in the storage media. In this manner the storage volume is considerable less than with systems employing chilled water storage.

When designing a system with a phase change material, the engineer should consider the difficulty in accomplishing a large tonnage discharge in the last hour or two of load satisfaction. The LMTD of the heat exchange media should be carefully calculated. It is a rare occasion that the actual ton/hours of the storage media would not be greater than the calculated ton/hour load. It is generally risky to take credit for the sensible cooling stored as the bulk of the usable cooling is stored within the phase change media as latent cooling.

d) PHASE CHANGE SYSTEMS - EUTECTIC SALTS

TES systems using rectangular plastic containers containing eutectic salts take advantage of the phase change which occurs during the melting of these salts. Not only is a certain minimal amount of sensible cooling available, but a considerable amount of cooling from the latent heat of fusion takes place during the phase change from solid to liquid.

One of the major advantages of such a system is that cooling can be stored using chilled water produced with a standard commercial water chiller. Because the eutectic salts can be frozen at close to normal chilled water temperatures, the higher refrigerant suction temperatures encountered allows the compressor to operate with less power consumption than would be used if ice were being generated.

When the salts are contained in rectangular plastic containers, they must be carefully stacked and baffled to control water flow through the layers of containers which is a labor intensive operation. Because the eutectic salts are more dense than water, and therefore non-buoyant, once they are placed in position they remain in place.

During the discharge cycle, the return water from the cooling coils is allowed to flow over and between the containers allowing the salts within the containers to change phase and absorb heat from the water. The method of sealing the containers must be carefully evaluated as screw type lids have been known to loosen thus allowing the salts to be contaminated with the surrounding water.

Systems using plastic spheres, filled with eutectic salts, can have the containers poured directly into the storage tank and because of their geometric shape the water flow can be passed through the bed of spheres without the use of baffles and hand placement of the containers. One major manufacturer of TES systems is currently developing such a product.

The storage volume of TES systems using containers of eutectic salts is generally considerably larger and therefore more costly than when ice is the storage medium. The selection of cooling coils within the air handling equipment is critical when using this type of TES system as the water temperatures are somewhat higher than conventional chilled water systems.

Coil area, rows and fin quantities must often be increased to compensate for these higher temperatures. The electrical savings at the chiller must be evaluated against the increased fan and pump horsepowers and the higher capital investment required.

These units are no longer in favor (1999) because of container deterioration, the higher water temperature and the higher capital investment.

e) PHASE CHANGE SYSTEMS - ICE, GENERAL

Ice as media for Thermal Energy Storage has been used for many years. Because of the relatively low refrigerant suction temperatures encountered when generating ice, the selection of the type of TES system must be carefully evaluated. The selection of refrigeration equipment is also important to receive the benefit of low operating cost during the ice generating mode.

One of the primary advantages of using ice as the storage media is that the water leaving the storage tank can be much lower than with other systems. Because a larger water temperature differential between supply and return water can be used, often as high as 25 degrees, as opposed to the more traditional 10 degrees. The circulating pump horsepowers and flow rates as well as pipe sizes can often be substantially be reduced.

The cooling coils are often smaller and the possibility of lower supply air temperatures can substantially reduce the air flow required with an attendant reduction in duct size as well as fan size and horsepower

f) PHASE CHANGE SYSTEMS - ICE ON COIL

TES systems where water flows over a chilled surface and which allows ice to build to a predetermined thickness is commonly encountered and often referred to as an "Ice Builder". The cold surface generally is a metal or plastic pipe or tube with a cold brine or refrigerant circulating within.

One drawback to allowing ice to build up on such a surface is that ice makes an excellent insulating medium and the temperature of the brine or refrigerant, and therefore the attendant refrigerant suction temperatures, must be considerably lower than some other TES systems. The thicker the layer of ice, the lower the suction temperature must be to offset the insulating effect of the ice.

This means an increase in the electrical power consumption during the ice building mode. This type of system is often the lowest cost from a capital investment standpoint but the lower cost must be evaluated against the higher operating cost.

Such "Ice Builder" systems are available in several different configurations. Modular system are sometimes used where the brine tubing is coiled and placed within insulated cylindrical plastic tanks or assemblies suspended within insulated rectangular tanks.

Other such systems use relatively large underground or above grade insulated tanks in which preassembled piping assemblies are placed. These piping assemblies may have brine circulating through the piping or have refrigerant contained directly within. During the discharge cycle, the chilled water returning from the cooling coils melts the ice which absorbs the heat from the water as phase change takes place. Such systems often use an air system where air is bubbled through the tank to agitate the water thus enhancing ice formation.

d) PHASE CHANGE SYSTEM - ICE HARVESTERS

A modification to the "Ice Builder" concept is the "Ice Harvester" TES system. This system is similar inasmuch as ice is allowed to build up on a cold surface. The problem with low suction temperatures encountered when a mass of ice is allowed to build which insulates the cold surface from the water being frozen, is minimized, although not entirely eliminated, with the "Ice Harvester".

The most common type of "Ice Harvester" allows ice to build up to a predetermined layer on evaporator plates or within evaporator tubes. Hot refrigerant gas is periodically passed into the evaporator to detach the sheet of ice attached to the evaporator which then drops into a tank below the "Harvester". The ice thus produced is referred to as fragmentary ice. The sheet ice falling into the tank shatters and becomes a pile of ice particles. Ideally during the discharge cycle, water is sprayed over this pile of ice and the melting material approaches 32 degree chilled water that is delivered to the load.

In applying this type of "Ice Harvester" system, several factors must be considered. The parasitic energy loss when hot refrigerant gasses are used to defrost the evaporator can be as high as 20% of the total energy consumed. The pile of fragmentary ice is deposited in a cone shaped deposit with an angle of repose of approximately 30 degrees. Because of this phenomena, a considerable portion of the tank is not effectively utilized and therefore a deep tank is required.

When the height of this cone of fragmentary ice is allowed to exceed 12 feet, the lower portion of the cone is subjected to considerable pressure causing heat to generate and therefore melts the ice and it re-freezes into a massive block of ice. The use of large tonnage "Ice Harvesters" should be evaluated very closely prior to selection.

The 1990 ASHRAE Refrigeration Handbook, Chapter 33.4 covers this problem in detail. In addition, even when a smaller cone of fragmentary ice is formed, unless the entire bed of ice particles is completely discharged each cooling cycle, a large block of ice sometimes forms and the advantage of the heat transfer efficiency of the water passing through a bed of ice particles is lost.

Such systems often have a dedicated refrigeration plant included as a part of the harvester package. This system is relatively complex and the equipment should be carefully evaluated prior to selection.

Another type of ice building system, which might loosely be described as an "Ice Harvester", is a system where a refrigerant, or chilled brine, passes through a large diameter drum. Depending on whether the freezing surface is on the interior or the exterior, the drum rotates and a cutting blade removes a thin sheet of ice from the exterior of the drum or the drum is stationary and cutters remove the ice from the interior surface of the drum. These systems are generally limited to process applications and are of relatively low capacity.

e) PHASE CHANGE SYSTEMS - SLUSH ICE

The use of ice in the form of slush is sometimes used for TES applications. This type of system uses a binary solution which partially freezes into a slush within a freezer which is generally mounted in a vertical position. The slush ice is collected in a storage tank and during the discharge cycle, the warm solution returning from the cooling coils is pumped to the upper part of the storage tank and allowed to percolate down through the mass of stored slush ice where the heat is absorbed.

Such systems are capable of delivering the lowest temperature of chilled solution of any of the other TES systems. The heat exchange during the discharge cycle is probably the most efficient of all the TES systems. This is a rather complex system and is relatively costly.

f) PHASE CHANGE SYSTEMS - ENCAPSULATED ICE

The remaining method of using ice as the storage media in TES systems is referred to as "Encapsulated Ice". This method comprises deionized water contained within a sealed plastic container either spherical, tubular or rectangular in shape. A nucleating agent within the sealed container is sometimes used and allows the water to change phase from liquid into solid almost instantaneously at 32 degrees. A brine chiller is used for cooling either a glycol or chloride solution which then flows around the containers freezing the ice contained within.

The brine is never in contact with the ice and during the discharge cycle is either circulated directly through cooling coils or through a plate type heat exchanger located in the chilled water loop. The energy consumed by the brine chiller in "Encapsulated Ice" systems is generally less than with any other ice based TES systems.

Systems using rectangular configured containers must have these containers placed in a tank very carefully. The tank dimensions and shape must be tailored to the dimensions of the containers. Baffles are placed between layers of stacked containers to provide proper flow circuiting and velocities through the bed of containers. Because ice is less dense than water the containers have a tendency to float which must not be permitted. To do so will allow a channeling effect and therefore reduce the amount of heat transfer surface available for load satisfaction.

The spherical shaped containers or "Ice Balls" as they are commonly referred to, may be placed in a tank of any configuration or size. These balls are allowed to float and therefore seek their own level. The brine passes vertically through the bed of balls with virtually non measurable velocity. Because of their geometric configuration, the fluid passage ways between the balls remains constant. The storage volume of such systems is the least of all of the TES systems.

Tubular shaped containers are placed vertically in a tank ant tightly packed together. An air space is provided within the tube to allow for expansion of the fluid when the liquid turns into ice. The brine passes vertically between the tubes at a very low velocity causing the water within the tubes to freeze solid. The storage volume of tubular configured systems is close to that of the spherical container systems.

"Encapsulated Ice" systems have the advantage of storing unused ice from one discharge cycle to another without the ice build up and bridging problems encountered with some other type ice storage systems.

REFRIGERATION EQUIPMENT SELECTION

The selection of refrigeration equipment is important in all TES applications. Chilled water and eutectic salt based storage systems can use water chillers with any type of the commonly available compressor. When ice is being generated, relatively low suction temperatures and high compression ratios are encountered. This situation calls for a careful evaluation of the type of compressor to be used.

The centrifugal compressors found on most moderate to large tonnage chillers are generally unsatisfactory for the operating conditions being encountered. Centrifugal compressors are efficient, but only within a narrow band of operation and in most cases the production of ice falls outside the efficient portion of the compressor "map".

Positive displacement compressors such as those of the rotary screw or reciprocating type are suitable for high compression ratio applications and are therefore the obviously choice for ice generation.

Because TES systems operate at "Off Peak" hours, the ambient dry bulb and wet bulb temperatures are relatively low. This allows use of lower condensing temperatures than would be encountered during day time hours. The lower the condensing temperature the lower the operating kWH. The limiting factor for low condensing temperatures is the method used for feeding refrigerant into the evaporator.

The thermostatic expansion valve most commonly encountered on chillers requires approximately 100 psig differential between the suction and condensing temperature to maintain the required mass flow. Electronic expansion valves or floats allow a pressure differential as small as 15 psig. Such a condition is a highly desirable feature in TES applications. A liquid line booster pump will also allow the condensing pressure to approach that of the suction as the pump mechanically provides the pressure to feed the evaporator.

Evaporative condensers are generally the best choice for load leveling applications where the compressors will operate during day time hours as well as during "Off Peak" conditions. Water cooled condensers are also a suitable selection.

Air cooled condensers are an excellent choice for systems that only operate during "Off Peak" hours. Because the ambient dry bulb temperature swings between day and night are much greater than corresponding wet bulb temperature swings, the air cooled condenser is the ideal choice unless physical size dictates otherwise.

When selecting a chiller or condensing unit for ice generating applications, standard commercial duty equipment should be avoided. Industrial duty equipment should be used for such applications. To meet environmental constraints, CFC refrigerants should be avoided. Some of the newer HFC or HCFC refrigerants appear to be a good choice. Ammonia is an excellent choice as a refrigerant for most ice making applications but may encounter resistance if the equipment is in close proximity to inhabited areas.

Chapter 5

THE COMPRESSION REFRIGERATION CYCLE

Refrigeration Engineers are well aware that the capacity of a refrigeration system is based on four (4) different factors.

1) Condensing temperature

2) Suction temperature

3) Refrigerant mass flow

4) Liquid refrigerant sub-cooling.

For a given mass flow (determined by the swept volume of the compressor) the higher the suction temperature and the lower the condensing temperature, the greater amount of liquid sub cooling produces the greater the system capacity. Obviously there are practical limits to these temperatures. For the purpose of this paper, our discussion will be limited to condensing temperature and the methods of condensing the refrigerant.

There are two (2) functions the condenser must perform.

1) De-superheating, condensing (phase change)

2) Sub-cooling.

The compressor must raise the temperature of the

refrigerant to a point where condensing can take place. The temperature of the refrigerant gas must be higher than the condensing medium. With an air cooled condenser at 95 degree ambient, the refrigerant temperature must be higher to accomplish any heat transfer.

In reviewing a pressure / enthalpy chart for a given refrigerant you will note there is a corresponding temperature for each pressure. This is based on the refrigerant being in a saturated condition. During the compression process not only does the saturated temperature increase in proportion to the pressure. During this function another phenomena takes place. The heat of the refrigerant gas is increased by the work of the compressor (kW) and is termed superheat. This causes the refrigerant temperature to be increased well above the equivalent saturated temperature. This superheat must be removed from the refrigerant gas before any condensing can take place.

After the refrigerant gas is condensed into a liquid some type of sub-cooling should take place. Just as superheat refers to the increase of refrigerant temperature above it's corresponding saturation temperature, sub-cooling lowers the refrigerant temperature below it's corresponding saturation temperature. Sub-cooling is required to maintain the refrigerant in liquid form between the condenser and the expansion point where the refrigerant is flashed into a vapor within the evaporator.

Refrigerant "flash gas" refers to the boiling of refrigerant within the liquid line. This reduces the mass flow capacity of the expansion valve and "starves" the evaporator thereby reducing system capacity. The appearance of bubbles within a liquid line sight glass does not necessarily mean the system is low on refrigerant and is often "flash gas" caused by insufficient sub cooling

We now see that the condenser must perform three (3) functions:

1) De-superheating

2) Condensing

3) Sub-cooling

The question now arises as to what type of condensers are available. Without considering the different sub-categories of condensers, there are just three (3) basic types of refrigerant condensers.

1) Air Cooled

2) Water Cooled

3) Evaporative Cooled

AIR COOLED CONDENSERS

With this type of condensing, air is drawn or blown through a finned coil. The flow rate is between 800 cfm and 1000 cfm per ton of refrigeration

Residential / light commercial systems almost always use this type of condensing. The reason this method is used is that the capital investment is less. Unfortunately in most cases the kW per ton is much higher than other methods. Remembering that the higher the condensing temperature, the greater the electrical consumption. Also the higher the condensing temperature, the less the refrigeration capacity. In considering package type air conditioning equipment there are two (2) major categories.

1) Residential / light commercial quality levels such as Carrier, Lennox, Rheem, etc.

2) Custom designed industrial / institutional levels such as Governair, Webco, Miller Picking etc.

The residential / commercial levels are lower cost equipment with mass production and high volume in mind. The condensers air generally sized for a condensing temperature 35 degrees higher than the ambient dry bulb conditions. This means that at 95 degree ambient the condensing temperature will be 130 degrees (approximately 1.25 kW / ton.) When the application is in a climatic area with an ambient dry bulb temperature of 110 degrees, the condensing temperature would be 145 degrees (approximately 1.5 kW / ton.)

The higher quality, and also more expensive, units can be selected with any size condenser. This is generally based on the condensing temperature being 25 degrees to 30 degrees higher than the ambient dry bulb temperature .In the case of the 25 degree approach this means that at 95 degree ambient the condensing temperature will be 120 degrees (approximately 1.15 kW / ton.) When the application is in a climatic area with an ambient dry bulb temperature of 110 degrees, the condensing temperature would be 135 degrees (approximately 1.3 kW / ton.)

WATER COOLED CONDENSERS

There are several types of water cooled condensers, "shell and tube" and "shell and coil." The former is the most commonly type in commercial refrigeration equipment such as water chillers. The hot superheated refrigerant gas enters the shell of the condensers and contacts the outer surfaces of tubes where water, generally from a cooling tower, passes through the tubes. The heat rejected passes to the cooling tower where it is rejected to the atmosphere as the water is cooled by evaporation.

Cooling tower efficiency is not based on the ambient dry bulb temperature but is based on ambient wet bulb temperature. With the same ambient wet bulb temperature, the cooling effect would be the same whether the ambient dry bulb temperature was 90 degrees or 120 degrees. This means that the water temperature to the condenser would be the same regardless of ambient dry bulb temperatures and the refrigeration capacity would not suffer when encountering the high ambient dry bulb temperature conditions.

Condensers on standard chillers are sized for condensing temperatures 20 degrees higher than the entering water conditions. The standard rating conditions for chillers is based on 85 degree water entering the condenser. This temperature equates to a condensing temperature of 105 degrees (.65 to 1.0 kW / ton).

High efficiency chillers have oversized condensers where the condensing temperature can be 15 degrees higher than the entering water conditions or 100 degrees (.5 to .55 kW / ton). This type of chiller general has an oversize evaporator which provides a higher suction temperature to assist in achieving the low kW / ton. There is a draw back to this type of efficiency when using centrifugal compressors however that is the subject of another paper.

EVAPORATIVE COOLED CONDENSERS

This type of condenser is generally the choice when low operating costs are the paramount factor. There is rarely a cold storage application where any other type of condenser is encountered. This type of condenser is often mistaken for a cooling tower. Only close observation will disclose that rather than water lines entering the devise, there are refrigerant line entering and leaving. A small spray pump is used in a closed loop arrangement.

These condensers are much more efficient than any other method of condensing because the superheated refrigerant gases pass directly to the evaporative condenser where water is sprayed directly on the exterior of the tubes and the air being forced over the tubes cools the water film to wet bulb temperature. The resulting condensing temperature is only 12 degrees to 15 degrees higher than the ambient wet bulb temperature. 90 degree condensing temperatures are a common selection point.

PARASITIC ELECTRICAL LOADS

The parasitic electrical load of an air conditioning system is made up of fans, cooling tower fans and pumps. In conducting a financial analysis it is necessary to look at the kW of these items in addition to the electrical consumption. Of the condensing methods described, the air cooled condenser fans have the highest parasitic load.

The cooling tower fan, on the water cooled system, has

a slightly greater electrical draw than the fan of the evaporative condenser. The overall advantage in parasitic loads goes to the evaporative condenser because the cooling tower pump, on the water cooled system, generally has the very close to same kW draw as the cooling tower fan. The spray pump on the evaporative condenser has between 25% and 30% of the kW consumption of the fan. This amount can be substantial when considering that the cooling tower pump and the spray pump operates during the time any compressor is in operation regardless of the amount of cooling load.

MAINTENANCE & SERVICE

There is often a misconception that the maintenance expense is less on air cooled condensing equipment than those which employ water. It is true that water treatment must be maintained on water cooled or evaporative condensing systems. When the wear and tear on the compressors using air cooled equipment is factored into the over all maintenance cost, the advantages goes to those systems having low condensing temperatures. It is rare to find a compressor on an air cooled unit that has operated over ten (10) years. The life expectancy is much less on a system having 24 hour per day operation. A compressor on a system with a water cooled or an evaporative cooled condenser can be expected to last from 18 to 30 years.

Annual maintenance on the three (3) condensing methods should be as follows:

AIR COOLED

1) Wash fin and tube surfaces annual. (Requires a high pressure washing unit)
2) Lubricate all bearings

3) Adjust and/or replace fan belts (on belt driven fans)

WATER COOLED UNITS

1) Check chemicals and feed pumps semi monthly

2) Remove heads from condenser and brush tubes (annually).

3) Replace gaskets on condenser heads (annually)

4) Eddy current testing on tubes (every 3 years)

5) Clean tower sump (annually)

6) Lubricate bearings (annually)

7) Adjust and / or replace belts (annually on belt driven fans)

EVAPORATIVE COOLED CONDENSER

1) Check chemicals and feed pumps semi monthly

2) Chemically clean exterior surface of tubes (annually)

3) Clean condenser sump (annually)

4) Lubricate bearings (annually)

5) Adjust and / or replace belts (annually on belt driven fans)

Annual maintenance on the air cooled condensing unit has a slight advantage over the evaporative cooled unit. The maintenance on the water cooled condensing unit is the highest of the three (3.)

Chapter 6

PREPARING YOUR FIRST BID

Over 90% of all construction work you do will be based on a fixed cost estimate. Don't let the word "estimate" fool you. When you submit a bid (actually a contract proposal) and you bid is "accepted" you have entered into a contract whether you realize it or not. 50 of the 51 states (Louisiana excepted) have adopted the Uniform Commercial Code as law. When you make an offer, either verbal or written, and it is accepted, there is a binding contract. You can see why the word "estimate" is a misnomer. Actually the estimate is what you believe it will cost to do the job. The final price on your bid is what your client expects to pay. It is therefore extremely important that you estimated costs are as accurate as possible. If you are submitting a "Budget Price" be sure it is clearly noted as such.

Contracting is a very competitive field with many knowledgeable competitors. If you find that you are getting most of the work you are bidding, slow down and take a good look to see if you are missing some items or are being to aggressive on your labor estimates.

Lets look at your first bid on the air conditioning of an existing residence. If you are dealing directly with the homeowner you will no doubt be the "prime contractor" meaning that you will be responsible for a "Turn Key" job including everything involved including roofing, painting cutting, patching, concrete or any other involved trade. You can always exclude any of these items but you should discuss this with the homeowner prior to submitting your bid. You will need to prepare a confidential work sheet to develop your costs in addition to the amount of profit you will be satisfied with.

PHASE	HOURS	MAT COST
(1) 5 Ton Package gas / electric unit	8.0	$1,225.00
(1) Cooling / heating thermostat	1.0	45.00
(9) Supply registers	4.5	88.00
(1) Return air filter grille	1.0	34.00
Covering and protecting furniture	3.0	10.00
Cutting and framing roof openings	4.0	15.00
Equip platform on roof	8.0	80.00
Fabricate roof flashings	1.5	15.00
Patch roof	4.5	45.00
Set unit in place with crane	By Crane Co	550.00
Cut openings in walls / ceilings for registers	12.0	20.00
Register boxes	3.0	35.00
Fabricate supply & return air plenums	2.0	55.00
Supply air ductwork	12.0	395.00
Return air ductwork	4.0	110.00
Gas line from meter to roof equip	8.0	65.00
Condensate drain	2.0	27.00
Thermostat wire	1.5	12.00
230 volt wiring from panel to unit	By Elect Contr	1,250.00
230 volt disconnect switch at unit	" "	w/ above
New breaker in sub feed panel	" "	w/ above
New electrical service if required	Not Required	0
Patch drywall	6.0	50.00
Paint drywall	3.0	25.00
Clean debris from site	8.0	0
Start up and test equipment	2.5	0
Instruct home owner on the operation	2.0	0
Building permit (if required)	.0	320.00
		$4.416.00

Labor direct cost with burden @ $39.00 x 101.5 hours	=	$3,958.50

Material cost and crane expense	=	4,416.00
Sales tax on material only (6.75%)	=	154.98
Service reserve (3%f above)	=	258.84
Total Direct Cost		$8.788.32

Overhead, General and Admin Expense (23%)	=	$2.021.31
Total Job Cost		$10,776.63

Anticipated Profit @ 15% of Total Job Cost	=	$1,612.49
Total Bid Price	=	**$12,393.12**

Pay no attention to the above amounts, hours or percentages as they are supplied to help you understand how to arrive at a bid amount. From this example you should now understand how to develop a simple bid price. Even though every job has different components and work phases, your estimate work sheet should be as detailed as possible so that you don't miss any major item.

Let's look at two of the most misunderstood items in establishing true costs. These items are Overhead, General and Administrative Expense and Payroll Costs.

OVERHEAD, G&A

From the "bean counter's" point of view these are two (2) separate items. From an estimators stand point they are one and the same. These are part of the costs of doing business that have not been itemized and accounted for on your estimating worksheet. In starting out a business some of the following items listed may not be required but as your business grows you need to incorporate the costs for these items because these costs must be covered before you start making a profit.

Accountants include workman's compensation

insurance as well as employer's payroll contributions. The writer prefers to incorporate these costs as part of the direct labor costs as the ratio between labor and material on jobs differ greatly and are easier to handle as a portion of labor.

OVERHEAD
Office rent
Electricity, gas, telephone
Computer software
Office equipment repairs
Office supplies
Equipment lease payments
Truck repairs and maintenance
Truck insurance
Liability insurance
Small tools (ladders, hand & power tools, extension cords, etc)
Supplies (screws, welding rod, silver solder, duct tape, etc)
Outside accounting and audits
Legal expense
Salesman bonuses
Health Insurance
Depreciation (your equipment will wear out and need replacement)

G & A EXPENCES
Estimator salary w/ payroll benefits
Engineer's salaries w/ payroll benefits
Salesman salaries w/ payroll benefits
Book keeper salary w/ payroll benefits
Secretary salaries w/ payroll benefits
Receptionist w/ payroll benefits
Officer's salaries w/ payroll benefits (This is your pay for your work)
Delivery truck driver w/ payroll benefits

LABOR COSTS (Those on Production)
Hourly pay scale
Union expenses (Health, welfare, pension, dues, etc)
State liability Insurance
State Unemployment
Federal employer costs
Any other costs directly related to hourly job employees

Any individual job will only support a given price. If they price exceeds that amount, whatever it may be, the job will not go forward. If we assume that your labor and material estimate is correct the only variable items are the overhead and profit. Let's further assume that a particular job will only support a bid price of $100,000. of this amount $65,000 is allocated for material and labor. This leaves $35,000 for overhead and profit.

Let's assume that your G & A and overhead will amount to $210,000 per year and your annual material purchases and labor projections would be (yes you may need a crystal ball) $700,000. This means that your overhead would $210,000 divided by $700,000 or 30%. If we go back and add 30% overhead to our job cost of $65,000 we arrive at an overhead figure of $19,500 bringing total job costs to $65,000 + $19,500 or $84,500. This would leave a projected profit of $100,000 - $84,500 or $15,500 or 18.3%. This would be a decent profit.

Before you go out and by that new ski boat lets assume that instead of the annual material and labor cost of $700,000 you had less sales and this amount was only $500,000. Your projected overhead of $210,000 would then be $210,000 divided by $500,000 or 42%. This would make our job look like this. $65,000 (labor and material) at 42% (overhead) would give you an overhead figure of

$27,300 or a total job cost of $65,000 + $27,300 or $92,300. This amount deducted from our $100,000 bid price would provide a very slim profit of $7,700 or 8.3%. Assuming that you don't want to consider a profit on costs of less than 15% you could structure your bid as follows. $65,000 x 1.42 x 1.15 = $106,145 bid price. If you could convince your customer to pay this amount, fine! If your competition's bid is $100,000 you have probably lost the job.

The solution is obvious. You must either reduce your overhead cost or increase your volume. This is where some soul searching is required as well as having a "heart to heart" conversation with your accountant.

Another thing about dealing with "bean counters." They talk about profits based on "margin," which is based on total sales and not on costs. Profit and Loss Statements (also referred to as Operating Statements) are based on a percentage of sales or "margin." Construction estimates are historically based on "mark up" and not "margin." Even though "margin" is theoretically the correct method, you should be aware of the difference so that you and your accountant are speaking the same language.

New construction is much more competitive than remodel work. Just as custom home construction is not as competitive as tract work or apartment construction. If your overhead is in line and you don't have the capital to increase your sales volume where you can compete, and still make a decent profit, you may wish to take a look at another market. This may be a wise decision , at least until your sales increase to the point where your overhead is down, percentage wise and where you are competitive in the market place where you wish to operate.

The service portion of the industry operates on traditionally on a "time and material" basis. This means that you charge for the material used on a service call plus a stated amount per hour for labor. Parts and material is traditionally marked up 100% and labor charged at a list price you have established. In southern California commercial service rates in 1999 were $60.00 to $65.00 per hour. Some service companies add a "truck" charge" which is an arbitrary figure and is generally used when a lower hourly rate is used.

On the face of things, the 100% mark up on parts and material coupled with a generous hourly rate seems like "fat city." All is not what it seems. The overhead on a service business is much higher than on construction as there is the infamous call back. You are dealing with complex machinery which is subject to many types of failure. Let's assume that you have just completed a service call where you changed a fan belt and filters. Before the end of the day the fan motor burns out. What do you think the customer is going to say? "It was OK until you worked on my unit, why should I have to pay you again for another service call?" Or possibly "If the motor was bad why didn't you change it while you were here?"

If this client is one of your better ones, you will probably have to go back and change out his motor charging him only for the motor and "eating" the labor and truck charge. At least this is true if you want to keep him as a customer. This is the reason a service company's overhead is high and they have to charge a higher rate for their work.

The above example also points out that when any component's condition is in doubt, change it out. The customer will be far happier than if you have to return later and change the defective component out. Even though the service call will be much more expensive for your customer, he will have a better opinion of you if everything then works fine and he doesn't have to call you back within a short period of time.

CHAPTER 7

CASH FLOW

Cash flow is the name of the game. You can have great profits on your jobs but if the money doesn't come in quick enough you could be out of business. To maximize cash flow there are a number of important things to remember.

1) You have to pay your material bills on time. If you don't your suppliers are not going to give you their best prices.

2) Your labor costs are going to have to be covered on payday.

3) You must make your Federal withholding deposits on time. Failure to do this can be the "kiss of death" to your new business.

4) Your rent and utility bills must be paid promptly.

5) Having profitable jobs is important but having your books showing large accounts receivable does not help you pay your bills

You are going to have to have an excellent relationship with your bank, your suppliers and your customers. As early as possible you need to establish a line of credit with your bank. As a new business owner this is not as easy as it sounds. Don't be surprised if the banker tells you to come back in two years with a fully audited financial statement and they will consider a line of credit. You sure don't need a banker like that. Before you establish a banking relationship, interview the bank manager and see if he (she) seems like a good match with you. Select a bank who does considerable business with contractors. Many bankers don't have the foggiest notion about construction and the methods of payment. When you establish a credit line expect to put up some type of collateral.

The type of project you take will govern your cash flow. Service work provides quick turn around on your money. If you are going to be doing residential service, the author suggests that you have a merchant account with some credit card company. It's much easier to collect the money, either a check or credit card, on the site when the home owners is happy to have his system operating. A few weeks later he may not be as appreciative.

If you plan on doing residential remodel type work, the money is generally pretty quick. Getting a small cash deposit from the owner helps cash flow. Be sure you check with your local State to see if this is legal and how large the percentage can be. It will also pay you to establish a "home improvement" loan arrangement with your bank.

If the home owner doesn't have the cash to pay for the installation you can then offer to finance the installation through the bank. Once you are set up as a dealer with the bank, you will then be like the car dealer who provides the financing at the time of sale. When the job is financed in this way, as soon as the job is complete, the home owner signs an acceptance form which you take to the bank and collect your funds. There isn't any quicker method of collecting your money.

The above two examples are those where you deal directly with the owner. When doing new construction there is generally a general contractor involved between you and the owner. More often than not these general contractors are not willing to pay there sub contractors until they collect their money from the owner.

When dealing with a builder/developer (one who is building the project on speculation with plans to sell the property after completion) you will find that he has taken out a "Construction Loan." These types of loans allow the developer to receive progress payments based on some prearranged schedule. Some tract housing developers work on a "5 draw" system. This means that there are 5 times during construction where the lender's inspector visits the job site to see if the work has been completed and will then authorize the lender to release a certain amount of money to the developer after which hopefully he will issue your check. On larger housing projects the funds are released twice a month.

Some developers use a method referred to as "Builder's Control." When this method is used a voucher is issued when you submit your invoice. You then take this voucher to the office of the "Builder's Control" and they will issue a check, assuming that the lender's inspector has approved the percentage of work. This method provides prompter payment than when the money flows through the builder himself.

If you are considering commercial or school work you will find that the payments are a little slower than those discussed previously. In the case of schools, there are many different individuals who must approve payments. The general contractor, the Architect/Engineer and the State inspector. After everyone approves the payment the school districts business manager presents the pay request to the school board at their next meeting. Sometimes during the summer months the board doesn't always have a quorum when members are on vacation. This means that the payment can be held up until a special meeting is called when a quorum is present. With school and commercial work, plan on waiting 60 days after your billing for payment.

The author mentioned earlier about establishing a good relationship with your bank and your suppliers. When payments get delayed it's important to be able to drop by your bank and have money placed in your account from your line of credit.

You might be able to work out some type of a joint check arrangement with your suppliers. This means that when you receive payment the check will be made out to you and your supplier. You then take the check to your supplier and they deduct the amount that is owed to them and release your portion directly to you. It is suggested that this arrangement be used for large dollar amount equipment shipments. When this arrangement is used and the payment from the general contractor is late, your supplier will put pressure on him rather than you.

When a general contractor "sets up" his job costs it's important for you to have an input. If you leave things entirely up to him, he may not distribute his costs properly and even though your work has been approved, there will be inadequate funds released for payment to you. Whether the project is a single custom home, a large housing project or a school, you must be pro-active in this matter. On housing work, your pay request breakdown should be incorporated with your bid along with exactly what you will be supplying and what you are excluding. Insist that your bid proposal be incorporated into your contract agreement as an "Exhibit." In this manner there will be no misunderstanding about payment schedules. A sample residential proposal is included as ATTACHMENT "A"

On schools and commercial work, be sure and submit a payment breakdown schedule as soon as you get word that you will be awarded the job. In this way the general contractor will have the information when he sets up his payment draw schedules and won't have to guess. A sample is included as ATTACHMENT "B."

On residential work you can easily exclude certain portions of the work which you don't wish to perform. On schools and commercial work the contracts will be awarded based on sections of the specification. When bidding this type of work carefully review the specifications as well as the plans and any notes thereon. Don't make assumptions. If it says you are to furnish something you better plan on doing so even if this is not your normal practice. When a specific brand of equipment is specified you better be sure that the equipment you plan on supplying will meet the consulting engineers approval. It will probably be wise to give him (her) a call and get his opinion.

Commercial and school work has something referred to as retention. This means that 10% of each of your billings will be held until the owner accepts the building. Unfortunately the entire project must be accepted not just your work. This means that a poor asphalt paving job can hold up your money until the owner is satisfied. This may not be fair but it is a fact of life.

Retention is not common on residential work so strenuously object if some builder or contractor tries to include retention. All he (she) is trying to do is work on your money to improve his (her) own cash flow.

In the Author's estimation the following item is one of the most important in maintaining good cash flow. This well known, but seldom admitted, method is referred to as "front loading." The name speaks for itself. This means that when you establish your payment draw breakdowns, include as much money as possible in the draws near the front end of the project. This means that once you get your first draw, you will be working on someone else's money.

A word of caution, when you have made payroll, paid your suppliers and see a nice cash balance in your bank account, don't get carried away and spend it. You'll need the funds to finish the job. The safest way to handle this is to see what the bare cost will be to complete the project and be sure you have that amount undrawn and remaining in the builders account.

Because of historically late payments, the only advantage that the Author sees in commercial, schools and other types of public works projects is that the money is sure. Generally a performance and payment bond is required to be provided to the owner by the general contractor. If he goes broke or skips out to Mexico, with all of the sub contractor's payments, the insurance company will step in and complete the job and see that every one gets paid.

The same is not true with the builder/developer. Even though the payments are generally quicker, there are no performance or payment bonds to protect the sub contractors and suppliers. For this reason three items are important to remember:

1) Check out the payment history of the developer. His Dun and Bradstreet report may look great but he may have been paying his suppliers and not his subcontractors. This would make his credit report look great when he actually is a deadbeat.

Check with other subcontractors who have been doing work for this developer. Particularly the trades who are on the project near completion. This includes the plumber and the concrete flat work contractor. An under capitalized builder may run out of money before the project is completed and the trades who must perform near the end of the project may have trouble getting paid. Don't forget that you will be setting equipment near the end of the project also

2) Front load the job. Work on his money and not yours whenever possible. If he goes broke, you will be holding his money and not be worried about yours. Any sophisticated builder is aware that his sub contractors will try to employ "front loading." You should try and break the payment phases down as fine as possible so that each phase has a relatively small percentage. The general contractor probably wont balk at a number of small percentage items but may if he sees a large percentage item near the beginning of the project.

Don't forget that if your project is subject to retention, increase each phase by at least the amount of retention so you have enough draw to cover your direct costa as well as your overhead and G & A expenses.

3) If you hear a rumor that some of the sub contractors are not being paid, let that be a warning flag. Communicate with the other sub contractors and find out

how things are going. You can learn many things using this method. When your payments are not being made when they are due, talk to the builder. As long as your work is up to date and there are no disputes on going don't hesitate to pull most of your crew from the job. Be sure and talk to your attorney at this stage.

TABLE 1

ROUND DUCT SIZING CFM

Residential Round Size	Residential Main	Residential Branch Duct	Residential SA Main	Commercial SA Branch	Commercial RA Main or Branch
5"	120	65	140	65	65
6"	150	120	250	120	120
7"	250	160	310	160	160
8"	310	210	440	210	210
9"	400	270	500	270	270
10"	500	320	600	380	380
12"	725	500	900	600	600
14"	1,000	650	1,000	900	900
16"	1,300	850	1,500	1,300	1,300
18"	1,600	1,100	2,000	1,800	1,800
20"	2,000	1,300	2,600	2,400	2,400
24"	2,800	1,900	4,100	3,900	3,900
30"	4,400	3,000	8,000	7,000	7,000
36"	6,500	4,300	13,500	11,200	11,200
42"	9,000	5,800	19,000	17,000	17,000
48"	12,000	7,500	27,000	24,000	24,000
54"	14,000	9,500	36,000	32,000	32,000
60"	18,000	13,000	49,000	42,000	42,000

TABLE 2

ROUND TO RECTANGULAR AIR DUCT CONVERSION

Round	Rectangular Equivalent					
5"	3 x 7	4 x 5				
6"	3 x 11	5 x 6				
7"	3 x 16	4 x 11	5 x 9	6 x 7		
8"	3 x 21	4 x 14	5 x 1	6 x 9	7 x 8	
9"	3 x 30	4 x 20	5 x 15	6 x 12	7 x 10	8 x 9
10"	3 x 40	4 x 25	5 x 19	6 x 15	7 x 12	8 x 11
12"	4 x 40	5 x 25	6 x 22	7 x 18	8 x 15	9 x 14
14"	5 x 42	6 x 32	7 x 26	8 x 22	9 x 19	10 x 17
16"	6 x 42	7 x 35	8 x 29	9 x 25	10 x 23	12 x 18
18"	7 x 46	8 x 39	9 x 33	10 x 30	12 x 23	14 x 20
20"	8 x 50	9 x 42	10 x 37	12 x 29	14 x 25	16 x 21
24"	10 x 55	12 x 43	14 x 35	16 x 30	18 x 26	20 x 24
30"	12 x 72	14 x 60	16 x 49	17 x 45	20 x 36	22 x 23
36"	15 x 85	17 x 72	18 x 65	20 x 56	24 x 41	30 x 36
42"	18 x 100	20 x 85	22 x 72	24 x 65	30 x 50	37 x 30
48"	22 x 100	25 x 85	30 x 65	34 x 68	36 x 55	38 x 49
54"	27 x 100	34 x 90	36 x 70	40 x 60	42 x 58	46 x 59
60"	33 x 100	35 x 90	38 x 80	42 x 72	48 x 63	52 x 56

TABLE 3

WATER PIPE SIZING

Sch 40 Pipe	GPM	Ft PD per 100'
1/2"	2	1.0
3/4"	3.5	3.2
1"	8	4.5
1 1/4"	16	4.1
1 1/2"	22	3.4
2"	40	3.0
2 1/2"	70	3.6
3"	130	3.9
4"	260	3.7
5"	460	3.5
6"	700	3.1
8"	1,500	3.3
10"	2,400	2.6
12"	3,500	2.2
14"	4,000	1.8
16"	6,000	2.0
18"	7,000	1.5
20"	8,000	1.1
24"	12,000	1.0

TABLE 4

GALLONS PER FT
CONTAINED IN PIPE

STND WT STEEL	GAL / FT	# / FT (Filled w/ Water)
1 /2"	.01	1.13
3/4"	.02	1.77
1"	.04	2.81
1 1/4"	.06	4.03
1 1 /2"	.09	5.26
2"	.16	8.17
2 1 /2"	.26	12.85
3"	.37	17.74
4"	.65	28.86
5"	1.02	42.85
6"	1.47	59.63
8"	2.61	100.83
10"	4.08	153.41
12"	5.88	212.19
14"	7.16	252.84
16"	9.49	325.22
18"	12.14	406.64
20"	16.73	541.70
24"	22.05	705.10

TABLE 5A

REFRIGERATION LINE SIZING (R22)

(Shown In Tons @ 40 F SST & 115 F SCT)

Size OD	Less than 100 Ft				101 Ft to 150 Ft			
	Suct Evap To Comp	Liq Rec To TXV	Disch Comp To Cond	Liq Drain Cond To Rec	Suct Evap To Comp	Liq Rec To TXV	Disch Comp To Cond	Liq Drain Cond To Rec
3/8"	NA	3.0	.7	1.5	NA	2.0	.5	1.5
1/2"	.6	3.5	1.2	1.8	.4	2.3	.8	1.8
5/8"	1.1	6.4	2.5	3.2	.7	4.3	1.7	3.2
7/8"	1.9	17.0	5.2	8.5	1.3	11.3	3.5	8.5
1 1/8"	5.8	34.4	11.8	17.2	3.9	22.9	7.9	17.2
1 3/8"	9.9	60.0	20.5	30.0	6.6	40.0	13.7	30.0
1 5/8"	15.9	95.0	31.9	47.5	10.6	63.3	21.2	47.5
2 1/8"	33.2	200.0	66.3	100.0	22.1	133.2	44.2	100.0
2 5/8"	58.1	354.0	115.6	177.0	38.7	235.8	77.0	177.0
3 1/8"	92.9	572.0	187.3	286.0	61.9	381.0	124.7	286.0
3 5/8"	139.5	800.0	281.0	400.0	92.9	532.8	187.1	400.0
4 1/8"	196.0	200.0	394.0	600.0	130.5	799.2	262.4	600.0
5 1/8"	355.0	NA	709.0	800.0	236.4	NA	472.0	800.0

Take precautions when suction lines are over 100 feet in length.

TABLE 5B

REFRIGERATION
LINE SIZING (R134a)

(Shown In Tons @ 40 F SST & 115 F SCT)

Less Than 100 Ft				101 Ft to 150 Ft			
Suct	Liq	Disch	Liq Drain	Suct	Liq	Disch	Liq Drain
Evap	Rec	Comp	Cond	Evap	Rec	Comp	Cond
To	To	To	To	To	To	To	To
OD	Comp	TXV	Cond	Rec	Comp	TXV	Cond	Rec
3/8"	NA	2.3	NA	1.2	NA	1.5	NA	1.2
1/2"	.7	2.8	.6	1.4	.2	1.9	.4	1.4
5/8"	1.1	5.3	1.1	2.6	.4	3.5	7	2.6
7/8"	1.8	14.0	2.9	7.0	1.2	9.3	2.0	7.0
1 1/8"	3.5	28.4	6.0	14.2	2.4	18.9	4.0	14.2
1 3/8"	6.2	50.0	10.4	25.0	4.1	33.3	6.9	25.0
1 5/8"	9.8	78.6	16.4	39.3	6.5	52.3	10.9	9.3
2 1/8"	20.2	163.0	34.0	81.5	13.5	108.6	22.6	1.5
2 5/8"	35.8	290.0	59.9	145.0	23.8	193.1	39.9	145.0
3 1/8"	57.1	462.0	95.5	231.0	38.0	458.2	63.6	31.0
3 5/8"	84.86	88.01	142.0	344.0	56.5	646.7	94.6	344.0
4 1/8"	119.4	971.0	200.0	485.0	79.5	NA	133.2	485.0
5 1/8"	213.0	NA	356.0	NA	141.9	NA	237.1	NA

Take precautions for oil return when suction lines are over 100 feet in length

TABLE 6

REFRIGERANT PRESSURE / TEMP CHART

TEMP F	R22 PSIG	R134a PSIG
-40	0.5	14.7*
-35	2.6	12.4*
-30	4.9	9.7*
-25	7.4	7.4*
-20	10.1	3.6
-15	13.3	0.0
-10	16.5	2.0
-5	20.3	4.2
0	24.0	6.5
5	28.2	9.1
10	32.8	11.9
15	37.7	15.1
20	43.0	18.4
25	48.8	22.0
30	54.9	26.1
35	61.5	30.4
40	68.5	35.1
45	76.0	40.1
50	84.0	45.5
55	92.6	51.3
60	100.6	57.3
65	111.2	64.1
70	121.4	71.2
75	132.2	78.7
80	143.6	86.8
85	155.7	95.3
90	168.4	104.4
95	181.8	114.0

*= Vacuum

TABLE 6 (Cont.)

REFRIGERANT PRESSURE / TEMP CHART

TEMP F	R22 PSIG	R134a PSIG
100	195.9	124.2
105	210.8	135.0
110	226.4	146.4
115	242.7	158.5
120	259.9	171.2
125	277.9	184.6
130	296.8	198.7
135	316.6	213.5
140	337.2	229.1
145	359.9	245.5
150	381.2	262.7
155	405.1	280.7

DENOTES INCHES OF VACUUM

TABLE 7

STEAM AND CONDENSATE RETURN LINE SIZING

# Per Hour @ 2% Pitch		SAT STEAM # Per Hour		
SCH 40 STEEL	COND RET*	@ 30 psig	@ 50 psig	@ 150 psig
1/2"	76	45	90	130
3/4"	161	100	200	300
1"	306	175	375	520
1 1/4"	635	270	550	1,150
1 1/2"	958	520	1,150	1,700
2	1,860	1,010	2,500	3,200
2 1/2"	3,000	1,700	3,200	5,500
3"	5,350	3,000	6,000	10,000
4"	11,000	6,000	14,000	20,000
5"	20,200	11,000	25,000	32,000
6"	32,900	19,000	35,000	55,000
8"	42,000	27,000	85,000	100,000
10"	80,000	58,000	100,000	NA
12"	100,000	100,000	NA	NA

* ALL CONDENSATE RETURN PIPING IS BASED ON GRAVITY FLOW.

TABLE 8

ELECTRIC MOTOR FULL LOAD AMPS

HP	115 v 1 Ph	230 v 1 Ph	230 v 3 Ph	460 v 3 Ph
.25	3.2	1.6	NA	NA
.333	4.6	2.3	.96	.48
.50	7.4	3.7	1.68	.84
.75	10.2	5.1	2.33	1.17
1	13.0	6.5	3.05	1.32
1.5	18.4	9.2	4.28	2.14
2	24.0	12.0	5.76	2.88
3	NA	17.0	8.28	4.15
5	NA	28.0	13.2	6.6
7.5	NA	40.0	19.3	9.7
10	NA	50.0	25.2	12.6
15	NA	NA	38.1	19.3
20	NA	NA	50.5	25.3
25	NA	NA	62.7	31.4
30	NA	NA	72.8	36.4
40	NA	NA	98.0	52.4
50	NA	NA	121	60.5
60	NA	NA	143	71.5
75	NA	NA	178	89.0
100	NA	NA	186	93.0
125	NA	NA	230	115
150	NA	NA	346	173
200	NA	NA	460	230

TABLE 9

ELECTRIC WIRE AMPACITY
AT TEMPERATURE RATINGS

(Per 100 foot run)

SIZE AWG or MCM	140 F Types T TW	167 F Types RH THW THWN XHHW THW	185 F Types V MI SA AVB THHN	194 F Types TA TBS	230 F Types AVA AVL	257 F Types AI AA
14	15	15	25	25*	30	30
12	20	20	30	30*	35	40
10	30	30	40	40*	45	50
8	40	45	50	50	60	65
6	55	65	70	70	80	85
4	70	85	90	90	105	115
3	80	100	105	105	120	130
2	95	115	120	120	135	145
1	110	130	140	149	160	170
0	125	150	155	155	190	299
00	145	175	185	185	215	230
000	165	200	210	210	245	265
0000	215	255	270	270	315	335
250	240	285	300	300	345	380
300	240	285	300	300	345	380
350	260	310	325	325	390	420
400	280	335	360	360	420	450
500	320	380	405	405	470	500

*The ampacity for type THHN for sizes 14, 12 and 10 are same as designated for THW.

TABLE 10

EVAPORATIVE COOLING SYSTEM SIZING

1) Establish Sensible Heat Load in building.
2) Establish 1% dry bulb and wet bulb design temps from ASHRAE manual.
3) Select efficiency of evaporative cooler (Usually between 80% and 85%)
4) Select discharge air temperature as follows:
   Ambient dry bulb - (Difference between ambient dry and wet bulb x eff%)
5) Establish inside design temperature which must be higher than discharge air.
6) Use the following formula to determine equipment size
   Sensible Load in Btu = Required CFM
   (Inside design temp-discharge air temp) x 1.08
7) See following example

A) 200 x 300 sq ft building
B) Sensible cooling load = 1,800,000 btu
C) Ambient Dry Bulb @ 1% = 103 degrees
D) Ambient wet bulb @ 1% = 67 degrees
E) Unit efficiency = 82%
F) Discharge air = 103-(103-67 x .82) = 73.5 degrees
G) Space design temperature = 79 degrees
H) Cfm Required = 1,800,000/(79.0 - 73.5) x 1.08 = 303,039 CFM

TABLE 11

HANDY FORMULAE

	Single Phase	**Three Phase**
Amps when HP is known	$\dfrac{Hp \times 746}{Volts \times Motor\ Eff \times Power\ Factor}$	$\dfrac{Hp \times 746}{1.73 \times Volts \times Motor\ Eff \times Power\ Factor}$
Amps when kW is known	$\dfrac{kW \times 1000}{Volts\text{-x-}Power\ Factor}$	$\dfrac{KW \times 1000}{1.73 \times Volts \times Power\ Factor}$
Amps when kVA is known	$\dfrac{kVA \times 1000}{Volts}$	$\dfrac{KVA \times 1000}{1.73 \times Volts}$
Kilowatts	$\dfrac{Volts \times Amps \times Power\ Factor}{1000}$	$\dfrac{1.73 \times Volts \times Amps \times Power\ Factor}{1000}$
kVA	$\dfrac{Volts \times Amps}{1000}$	$\dfrac{1.73 \times Volts \times Amps}{1000}$

TABLE 11
(Cont.)

Horsepower (Electric)	$\dfrac{\text{Volts} \times \text{Amps} \times \text{Motor Eff}}{746}$	$\dfrac{1.73 \times \text{Volts} \times \text{Amps} \times \text{Motor Eff} \times \text{Power Factor}}{746}$

HANDY FORMULAE

BTU (Water) GPM x 500 x Temp Diff	BTU (Sensible - Air) CFM x 1.08 x Temp Diff
Tons of Refrigeration $\dfrac{\text{BTU}}{12,000}$	BTU (Total - Air) CFM x 4.5 x Enthalpy Diff
Horsepower (Boiler) = 34.5 # steam @ 212	FHorsepower per Hour = 2.545 BTU
1 Psig = 2.31 Feet of Head (Water)	KW = 3,413 BTU
Degrees Celsius = (Degrees F – 32)/1.8	Degrees Fahrenheit = (1.8 x Degrees C) + 32

TABLE 12

ENTHALPY (H) VALUES OF DRY AIR

WET BULB F	H	WET BULB F	H	WET BULB F	H	WET BULB F	H
50.0 =	20.30	60.0 =	26.46	68.0 =	32.42	76.0 =	39.57
50.2 =	20.42	60.2 =	26.53	68.2 =	32.59	76.2 =	39.77
50.4 =	20.57	60.4 =	26.74	68.4 =	32.67	76.4 =	39.98
50.6 =	20.64	60.6 =	26.87	68.6 =	32.92	76.6 =	40.17
50.8 =	21.32	60.8 =	27.01	68.8 =	33.08	76.8 =	40.37
51.0 =	20.86	61.0 =	27.15	69.0 =	33.25	77.0 =	40.57
51.2 =	20.98	61.2 =	27.29	69.2 =	33.42	77.2 =	40.77
51.4 =	21.09	61.4 =	27.43	69.4 =	33.59	77.4 =	40.97
51.6 =	21.21	61.6 =	27.57	69.6 =	33.75	77.6 =	41.18
51.8 =	21.32	61.8 =	27.71	69.8 =	33.92	77.8 =	41.38
52.0 =	21.44	62.0 =	27.85	70.0 =	34.09	77.0 =	41.58
52.2 =	21.55	62.2 =	27.99	70.2 =	34.26	78.2 =	41.79
52.4 =	21.67	62.4 =	28.14	70.4 =	34.43	78.4 =	42.00
52.6 =	21.79	62.6 =	28.28	70.6 =	34.52	78.6 =	42.10
52.8 =	21.90	62.8 =	28.43	70.8 =	34.79	78.8 =	42.31
53.0 =	22.02	63.0 =	28.57	71.0 =	34.95	79.0 =	42.62
53.2 =	22.14	63.2 =	28.72	71.2 =	35.13	79.2 =	42.83
53.4 =	22.24	63.4 =	28.87	71.4 =	35.30	79.4 =	43.05
53.6 =	22.38	63.6 =	28.94	71.6 =	35.48	79.6 =	43.26
53.8 =	22.50	63.8 =	29.15	71.8 =	35.65	79.8 =	43.48
54.0 =	22.61	64.0 =	29.31	72.0 =	35.83	80.0 =	43.69
54.2 =	22.74	64.2 =	29.46	72.2 =	36.01	80.2 =	43.91
54.4 =	22.86	64.4 =	29.61	72.4 =	36.19	80.4 =	44.14
54.6 =	22.98	64.6 =	29.76	72.6 =	36.38	80.6 =	44.34
54.8 =	23.10	64.8 =	29.91	72.8 =	36.56	80.8 =	44.56
55.0 =	23.22	65.0 =	30.06	73.0 =	36.74	81.0 =	44.70
55.2 =	23.34	65.2 =	30.10	73.2 =	36.92	81.2 =	45.00
55.4 =	23.47	65.4 =	30.37	73.4 =	37.11	81.4 =	45.23
55.6 =	23.59	65.6 =	30.52	73.6 =	37.29	81.6 =	45.45
55.8 =	23.72	65.8 =	30.68	73.8 =	37.48	81.8 =	45.69
56.0 =	23.84	66.0 =	30.83	74.0 =	37.66	82.0 =	45.90
56.2 =	23.97	66.2 =	30.99	74.2 =	37.85	82.2 =	46.13
56.4 =	24.10	66.4 =	31.15	74.4 =	38.04	82.4 =	46.36
56.6 =	24.22	66.6 =	31.30	74.6 =	38.23	82.6 =	46.58

TABLE 12 (Cont>)

ENTHALPY (H) VALUES OF DRY AIR

WET BULB F	H	WET BULB F	H	WET BULB F	H	WET BULB F	H
56.8 =	24.35	66.8 =	31.46	74.8 =	38.42	82.8 =	46.81
57.0 =	24.48	67.0 =	31.62	75.0 =	38.61	83.0 =	47.04
57.2 =	24.61	67.2 =	31.78	75.2 =	38.80	83.2 =	47.28
57.4 =	24.74	67.4 =	31.94	75.4 =	39.00	83.4 =	47.51
57.6 =	24.86	67.6 =	32.10	75.6 =	39.19	83.6 =	47.75
57.8 =	24.99	67.8 =	32.26	75.8 =	39.38	83.8 =	47.98

Enthalpy of air (h) is the amount of btus in 1# of dry air at a given wet bulb temperature.

TABLE 13

COOLING LOAD CHECK FIGURES

These figures do not replace the need for running an actual cooling load on the building but are useful for budgeting and rough analyses.

Type Structure	Sq Feet Per Occupant	Sq Feet Per Ton
Apartments (High Rise)	175	400
Auditoriums	11	200
Churches	11	200
Department Stores (Basements)	25	285
Department Stores (Street Level)	25	245
Department Stores (Upper Stories)	55	340
Factories (Assembly Areas)	35	150
Factories (Light Manufacturing)	150	150
Factories (Heavy Manufacturing)	250	80
Hotels & Motel Rooms	150	300
Hospitals (Patient Areas)	50	220
Hospitals (Public Areas)	80	140
Libraries	60	280
Malls	75	230
Museums	60	280
Residences (Large)	400	500
Residences (Medium)	360	550
Restaurants (Large)	15	120
Restaurants (Medium)	15	100
School & College Classrooms	25	200
School Gymnasiums	11	150
Theaters	11	200

ATTACHMENT "A"

SAMPLE RESIDENTIAL BID

ACME AIR CONDITIONING
21450 Center Ave
Redlands, California 92324
(909) 999-1234

October 11, 1998

Meadowlark Homes
1234 Magnolia Blvd
Redlands, California 92375

Atten: Sam Smith

Gentlemen:

We propose to furnish and install a complete heating and air conditioning system in the new house at 203 Citrus Ave, Redlands. Work will be performed in accordance with plans and specifications to prepared by our Engineering Department and in accordance with local codes as follows.

Rheem AC-36-100 Gas Electric Unit on roof
Night Set Back Thermostat with control wiring
Return Air Filter/Grille in Hallway Ceiling
PVC Condensate Drain
9 Supply outlets to bedrooms, kitchen. living room, dining room and exterior bath

Vents for bathroom fan and range hood
1 year parts and labor warranty, 4 years extended warranty on compressor only

For the sum of $3,165.00

Exclusions: Cutting and framing for opening, range hood, bathroom fan, electrical hook up to unit, general sheet metal.

TERMS: NET

40% on Duct rough in, 55% on setting equipment, balance on setting registers & stat.

John Brown
License 12345678

ATTACHMENT "B"

SAMPLE JOB COST BREAKDOWN

ACME AIR CONDITIONING
21450 Center Ave
Redlands, California 92324
(909) 999-1234

July 4, 1776

BIG DAWG CONSTRUCTION
345 Industrial Place
Redlands, Ca 92374

Attention: Carl Carlson

RE: Sunlight Elementary School

Dear Carl:

Please incorporate the following cost breakdown in your payment schedule. If you have any questions, please give me a call.

Underground Ductwork for Multipurpose Bldg	10.0%
Bldg K Ductwork Rough In	11.0%
Multipurpose Bldg Duct Rough In	13.0%
Bldg K Registers and Grilles	4.0%
Multipurpose Bldg Registers and Grilles	4.0%
Bldg K AC Units on Roof	24.0%
Multipurpose Bldg AC Units on Roof	25.0%
Start Up	5.0%
Air Balance	4.0%

Best regards,
John Brown

About the Author

With a Masters Degree in Mechanical Engineering and over 45 years experience as a consulting engineer, an equipment vendor and a mechanical contractor, Delbert D. Thomas is well qualified to advise the individual who is considering entering the HVAC business as an owner.

He has been active in the American Society of Heating, Refrigerating and Air Conditioning Engineers (ASHRAE) and is Past president of a local chapter and has served as a Regional Vice Chair (Refrigeration) for the Society. Mr. Thomas has been published in many different national industry publications and is an internationally acknowledged authority on Thermal Energy Storage, Turbine Inlet Air Cooling and District Heating and Cooling Systems.

He currently operates a mechanical consulting firm in southern California known as Mechanical Systems. Mr. Thomas also is active in marketing simple Do-it-yourself HVAC systems and gold mining equipment designs on the Internet through Poorboydesigns.com.

www.ingramcontent.com/pod-product-compliance
Lightning Source LLC
Chambersburg PA
CBHW030744180526
45163CB00003B/911